The Fifth Hammer

The Pythagorean Forge (anonymous eleventh-century illustration, Barbara Münxelaus, *Pythagoras Musicus* [Bonn, 1976], fig. 11).

The Fifth Hammer

Pythagoras and the
Disharmony of the World

Daniel Heller-Roazen

ZONE BOOKS · NEW YORK

2011

ZONE BOOKS
1226 Prospect Avenue
Brooklyn, NY 11218

Printed in the United States of America.

Distributed by The MIT Press,
Cambridge, Massachusetts, and London, England

Library of Congress Cataloging-in-Publication Data

Heller-Roazen, Daniel.
    The fifth hammer : Pythagoras and the disharmony of
the world / Daniel Heller-Roazen.
        p. cm.
    Includes bibliographical references and index.
    ISBN 978-1-935408-16-1
        1. Pythagoras. 2. Music—Philosophy and aesthet-
ics. 3. Music theory—History. 4 Cosmology. I. Title. II.
Title: 5th hammer.

    ML3800.H333 2011
    781.2'501–DC22

                            2011001453

«OCTAVE», s. f. La première des consonances dans l'ordre de leur génération. L'octave est la plus parfaite des consonances ; telle est, après l'unisson, celui de tous les accords dont le rapport est le plus simple : l'unisson est en raison d'égalité, c'est-à-dire comme 1 est à 1: l'octave est en raison double, c'est-à-dire comme 1 est à 2 ; les harmoniques des deux sons dans l'un et dans l'autre s'accordent tous sans exception, ce qui n'a lieu dans aucun autre intervalle. Enfin ces deux accords ont tant de conformité, qu'ils se confondent souvent dans la mélodie, et que dans l'harmonie même on les prend presque indifféremment l'un pour l'autre.

Cet intervalle s'appelle *octave*, parce que, pour marcher diatoniquement d'un de ces termes à l'autre, il faut passer par sept degrés et faire entendre huit sons différens.

—Jean-Jacques Rousseau, *Dictionnaire de musique* (1768)

OCTAVE. The first of the consonances in the order of their generation. The octave is the most perfect of the consonances; it is, after the unison, that of all the concords whose connection is the most simple; the union is in the computation of equality, that is, as 1 to 1. The octave is in double computation, that is, as 1 to 2. The harmonies of the two sounds agree in each without exception, which has no place in any other interval. Lastly, these two concords have so much conformity, that they are often confounded in melody, and in harmony itself we take them indifferent one from the other.

This interval is called octave, because, to move diatonically from one of thee terms to the other, we must pass by seven degrees, and make eight different sounds be heard.

—*A Complete Dictionary of Music, Consisting of a Copious Explanation of All Words Necessary to a True Knowledge and Understanding of Music, Translated from the original French of J. J. Rousseau by William Waring* (1779)

Contents

Preface

According to a long tradition, Pythagoras was the inventor of harmony, understood in a double sense: as an account of a limited set of musical sounds and, more broadly, as a doctrine of the intelligibility of the natural world. This book investigates both aspects of the invention attributed to Pythagoras. It seeks to show how, from the ancient to the medieval and the modern period, the analysis of sounds in terms of quantities furnishes a model for cosmological inquiry. That inquiry, which may have constituted the first example of science as we know it, rests on a simple practice: a transcription of the world in the units of mathematics. In this sense, the Pythagorean project is one of reading and notation, which aims to decipher and transcribe the signs written in the great and often abstruse book of nature. One may say that the notion on which such a practice of representation rests is the "letter," if one takes that term, in the old sense, as signifying a minimal element of intelligibility and if one adds that such minimal elements are quantitative in nature. The world may be deciphered if it can be resolved into such letters: this would be one way to express a long-lived Pythagorean wager.

From the pre-Socratic period through the Middle Ages and the epoch of modern science, Pythagorean notation, however, runs up against a limit. Something resists being recorded in any quantitative units, be they notes, numbers, lines, or figures. There are at least two fundamental reasons for the insistence of this limit. Letters, first of all, may well be inadequate to the whole world they aim to record. But it is also conceivable that the whole world cannot, in the end, be grasped as a whole. These two possibilities can be considered independently or in conjunction; moreover, depending on the author and the age, the meanings given to the "letter" and to the "world" may also vary. Disharmony comes in many types. Yet in two fundamentally distinct epistemological and metaphysical paradigms, before and after the break traditionally associated with Galilean science, thinkers who sought to order the natural world by means of elements of quantity encountered something that, while audible, would not allow itself to be recorded.

For reasons the reader will quickly learn, "the fifth hammer," in this book, names that unsettling part, which troubles Pythagorean musical theory and cosmology. Its differing yet obstinate recurrences constitute the matter of the coming chapters.

CHAPTER ONE

Into the Forge

Pythagoras knew not to trust his ears. Sage and scientist, he understood that such organs of perception might well reveal some truths, but he was aware that they might always also lead him into error. In his *Fundamentals of Music*, Boethius relates that Pythagoras "put no credence in the ears," since, being bodily parts like all others, they are subject to incessant change.[1] Sometimes they vary on account of external and accidental circumstances; sometimes they begin to differ by necessity, as when they age. Pythagoras could hardly have expected more from acoustical devices made by men. Musical instruments were, for him, "sources of much variability and inconstancy."[2] More than once, he had studied the nature of strings. Their tones may change for reasons almost too numerous to enumerate. Depending on their matter, depending on their length and width, depending on the air about them and the force with which one plucks them, cords will inevitably produce different noises. Pythagoras had observed that other instruments are of a like nature. The sole certainty about them is that sooner or later, they must abandon "the state of their previous stability," causing their tones to change perceptibly. In awareness of the implications of these facts, Pythagoras had resolved to liberate himself and his investigations, as much as he could, from the troubling consequences of sensible things. He would study the laws of sound without needing to encounter them in bodily form, and he would acquire knowledge of the properties of audible things by the aid of reason alone.

The project was boldly conceived, but it was never to reach

completion. Just as Pythagoras decided on the course his inquiry must take, he was suddenly distracted. "As if impelled by a kind of divine will," the thinker found himself led from the place of his customary calculations out into the world. Enchanted, he wandered toward a forge. From within the smithy came the noise of many hammers. "Somehow," we learn, "they emitted a single consonance from differing sounds." Astonished, Pythagoras began to grasp what he had discovered. "He was in the presence of what he had long sought, and he approached the smiths' work as if spellbound."[3]

Wonder soon passed to reasoned reflection. The noise was not autonomous; it came from the activity of men, carried out with tools in one particular surrounding. Pythagoras sought to learn the cause of the coincidence of sounds, but this was no easy task. How to tell the smiths from the smithy, the hammerers from the hammers? First he tested a hypothesis. "He commanded the workers to exchange their tools among each other." Then the blacksmiths began their work anew. The startlingly single consonance remained. At least one indubitable conclusion could now be drawn: "The property of sounds did not rest in the muscles of men; instead, it followed the exchanged hammers." Yet the concord, properly defined, could not be said to move. Its place was stable; indisputably, it lay not in the workers, but in their tools. More exactly, the consonance lay in one of the hammers' many sensible properties. This was a property that may well have been incidental to their instrumentality: namely, that of possessing mass, insofar as it may be measured with precision. Pythagoras was quick to recognize this point: the "single consonance" resulted from the relations between the hammers' weights, which caused a set of pleasing sounds. For Boethius, a thinker of late antiquity, the relations of the various hammers' weights were naturally best expressed in the technical terms of Graeco-Latin arithmetic. He writes:

> There happened to be five hammers, and those which sounded together the consonance of the diapason were found to be double in weight. Pythagoras determined further that the same one, the one that was

the double of the second, was the sesquitertian of another, with which it sounded a diatesseron. Then he found that this same, the duple of the above pair, formed the sesquialter ratio of still another, and that it joined with it in the consonance of the diapente. These two, to which the first double proved to be sesquitertian and sesquialter, were discovered in turn to hold the sesquioctave ratio between themselves.[4]

That summary, Boethius concedes, might also be more simply put. It suffices to imagine the following account: "So that what has been said might be clearer, let the weights of the four hammers be contained in the following numbers: twelve; nine; eight; six."[5]

Pythagoras immediately understood that the discovery was significant. Quickly, he returned home and, by new means, repeated the experiment. "First," Boethius narrates, "he attached corresponding weights to strings and discerned by ear their consonances; then, he applied the double and mean and fitted other ratios to lengths of pipes." Next, he "poured ladles of corresponding weights into glasses, and he struck these glasses—set in order according to various weights—with a rod of copper or iron." Finally, "he turned to strings, measuring their length and thickness, that he might test further."

He reached a set of findings that could be demonstrated most easily by means of a simple device: the monochord. This instrument consists of a single string, stretched over a sound box, fastened at both ends, and whose length is divided by a bridge that can be moved at will. When plucked or played, the cord will emit a single tone. It is obvious that as the length of the string is gradually diminished, its pitch will grow increasingly acute. But with the ring of the smithy still in his ears, Pythagoras had come to grasp far more. Regularities could be observed between adjustments in length and changes in sound; correlations could therefore be established between geometrical and sonorous phenomena. Three equivalences, in particular, were immediately remarkable. An open string will produce a tone. When it is halved, it will produce another, exactly one octave above the first. Then the cord will have emitted the interval known to the ancients as "diapason." When, instead, the string is divided into three sections, two of which are played, a new interval will be audible: the fifth, known to the Greeks and Romans as "diapente." When, finally, the string is divided into four sections, three of which are played, the cord will produce a tone higher than the open string by the distance of a fourth; thus the instrument will have sounded the "diatesseron."

Suddenly, the seemingly endless diversity of sounds acquired a new simplicity. Acoustical intervals were now expressible as arithmetical relations. The proof lay in the reduction of the sound of the octave to the relation of two to one (2:1); that of the fifth, to the relation of three to two (3:2); that of the fourth, to the relation of four to three (4:3). In short, the natural world could be transcribed—not, to be sure, by the letters of alphabets, which differ among themselves according to the variety of human idioms, but by those various collections of unity that the ancients understood to be "numbers." The consequences of this fact for the understanding of the physical world were great. In sensible things, one could find the intelligible; in the changing, the immutable. Through his analysis of sound, Pythagoras had reached the basis of his metaphysics. This was a doctrine Aristotle recorded in a series of lapidary propositions: "things are the same as numbers" (ἀριθμοὺς εἶναι αὐτὰ τὰ πράγματα);[6] "things are like numbers" (μιμήσει τὰ ὄντα φασὶν εἶναι τῶν ἀριθμῶν);[7] "things and numbers are composed of the same elements" (τὰ τῶν ἀριθμῶν στοιχεῖα τῶν ὄντων στοιχεῖα πάντων ὑπέλαβον εἶναι).[8] Such statements suggest differing, perhaps conflicting, positions.[9] But despite their variety, they attribute a single program to Pythagoras and his successors: to find, in the idea of number, a key to the understanding of the natural world.

That understanding could advance far, yet ultimately it was to falter. One may take the findings in the forge as an illustration of the project and its limits. Pythagoras developed an arithmetical doctrine founded on a series of four terms that corresponded to the four respective weights of the hammers in consonance: twelve;

nine; eight; and six. Restricting himself to these integers, Pythagoras could thus express the numeric relations that, on a monochord, produce the octave (12:6), the fifth (9:6, or 12:8), and the fourth (8:6, or 12:9). But those relations also reflected simpler proportions. "Twelve to six" might be rewritten as "two to one"; "nine to six, or twelve to eight" could be rewritten as "three to two"; "eight to six, or twelve to nine" might be rewritten as "four to three." In short, all three fundamental acoustic intervals were expressible by the relations of the first four natural numbers; these terms alone sufficed for the analysis of the concord in the forge. Later, Pythagoras's followers would take a further step. The first four numbers became, for them, the units of a cosmological "fourfold" (τετρακτύς).[10] Their arithmetical summation would produce the unity of ten, furnishing the basis of all further calculation. Their geometrical arrangement would compose the "pebble figure" of the perfect triangle—that is, an equilateral triangle each of whose sides contains four points, followed by three, two, and one. Thinkers in the tradition would attribute diverse and far-reaching values to each of these arithmetical elements. Speusippus, for example, taught that the point is one; the line two; the triangle three; the pyramid four.[11] A source indicates that, among Pythagoreans, the principle was the object of a sacred oath: "No, I swear by he who transmitted to our soul the fourfold, / Which contains the source and the root of eternal nature."[12]

Yet the truth is that the set of four was lacking from the beginning. In the forge, the transcription was obstinately incomplete. One remained uncounted. Describing the instruments that produced the concord of sound, Boethius observes: "There happened to be five hammers." The relative weights of all but one could be perfectly notated with the aid of the first four numbers. But there was also a fifth. Boethius devotes no more than one sentence to the fate of this most unmusical of instruments: "The fifth hammer," he writes in passing, "which was discordant with all the others, was discarded" (*Quintus vero est reiectus, qui cunctis erat inconsonans*).[13] Sudden, certain, and apparently irrevocable, that "discarding" merits some reflection. What was the fifth hammer, such that Pythagoras chose so decidedly to reject it? Boethius suggests only the most minimal of answers, and it is not easily understood. He writes that the fifth hammer was "discordant with all the others." But that statement implies a question: What is this "all," if something—if even only one thing—sounds in utter dissonance with it?

The presence of the fifth hammer seems to belie the totality of the set of four. It could do so, however, in at least two ways, which suggest differing and indeed contradictory interpretations of the last instrument of percussion. Boethius's public may have concluded that the presence of the fifth hammer betrayed a fault to be ascribed not to Pythagoras, but to our own lowly world. Such an explanation might well be reconciled with the canons of ancient knowledge. It might be recalled that the classical thinkers, as a rule, sought to grasp principles of nature that would by definition be eternal, immutable, and necessary, but they also taught that particular things are by essence corruptible, changing, and hence uncertain. The early readers of Boethius's work may have understood the fifth hammer in such terms: for them, its discordant sound may have testified to the limits of the sublunary sphere, where natural science, even in its most developed forms, could predict no physical event with certainty. Only beyond the moon, in the noble regions that enclose this corruptible world, could mathematical principles and deductions find an exact realm of application. But today, of course, there is a more obvious solution to the problem of the noisy part. One may choose to point an accusing finger at the primitive, if ingenious theorist, inferring simply that something in the Pythagorean calculus was amiss. An error—whether of observation or outlook, measurement or method—could also have kept Pythagoras from finding a place for the tool in his system of proportions. One might well reason that had his analysis been correct, it would have admitted no remainder, for a scientific study surely tolerates no exceptions. Yet each solution, however imaginable, conceals an obscurity. What was the world of ancient knowledge, if it allowed—and perhaps demanded—a sound discordant with "all the others"? And what is the universe of modern science, if it, by contrast, cannot permit the noise of a single inconsonant part?

One can only speculate as to the reasons for the fifth instrument's discord. But this much can hardly be contested: although Pythagoras wished not to include the last hammer in his equivalences of noise and number, he nonetheless perceived it. Boethius leaves little doubt: immobile before the forge, "as if spellbound," the sage "overheard the beating of hammers somehow emit a single consonance from differing sounds." Thus the fifth tool beat, no less than one of five. Perhaps, in his momentary distraction, Pythagoras found himself drawn to that very instrument: the hammer with no number and no master, which somehow—yet impossibly—sounded both "in a single consonance" and in utter discordance "with all." One wonders whether the "kind of divine will" that caused the thinker to abandon his sheltered contemplations may not have had a part to play in this mysterious quintet. The spirit that deterred Pythagoras from his reasoned inquiry may have also been the one that remitted him to the sensible organs that he never meant to trust. Doubtless, the dispossession was transitory. But its consequences were to be lasting. Pythagoras might well return to his research; he might well reject all instruments. Dimly or distinctly, if only for a moment, he had nonetheless perceived a being without measure. It is difficult to imagine that he thought nothing of it. Yet it is certain that his followers, Boethius not least, later took pains to record it. Transcribed as an event of sound to which no certain quantity could be assigned, that resonance was to lure others into the forge discovered by Pythagoras. There, in fidelity and infidelity to their master, they would learn to perceive the harmonies of a music that no numbers may transcribe.

CHAPTER TWO

Of Measured Multitude

That Boethius should have attributed the discovery of the ratios of consonances to Pythagoras may seem a curious fact today. In its own time, however, the late antique philosopher's choice was perfectly traditional. The Pythagoreans were renowned in the classical world for their research into music as well as mathematics, and many sources suggest that the Pythagoreans held the properties of sound to illustrate the principles of number. A good thousand years before Boethius, Socrates alluded in the *Republic* to the "Pythagoreans" who sought the arithmetical causes of audible consonances.[1] In the portion of the *Metaphysics* devoted to the doctrines of "the so-called Pythagoreans," Aristotle attributed to these early thinkers a similar teaching. "They applied themselves to mathematics," the Philosopher recounted,

> and were the first to develop this science, and through studying it they came to believe that its principles are the principles of everything. Since they saw that numbers are by nature first among these principles...and since they also saw that the properties and ratios of harmony are based on numbers, it seemed clear to them that all other things have their whole nature modeled upon numbers, and that numbers are the ultimate things in the whole physical universe.[2]

A fragment thought to derive from a lost Aristotelian treatise, *On the Pythagoreans*, is more explicit:

> Having devoted themselves to mathematics and having admired the accuracy of its arguments, because it alone, among human activities,

admits of proofs, they saw the facts of harmony, that they happen on account of numbers...and they deemed the facts of mathematics and their principles to be generally the cause of all existing things; whoever wishes to comprehend the nature of existing things should therefore turn his attention to these, that is, to numbers and proportions, because it is by them that everything is made clear.[3]

Classical sources suggest that Pythagoras lived in the sixth century B.C. It is reported that he was born on Samos and that he left Ionia to settle in a colony of Magna Graecia sometime around 530 B.C.[4] None of his works, if there were ever any, survive. Although there is evidence of the presence of Pythagorean teachings throughout southern Italy from the fifth century onward, texts by the earliest disciples are all lost. There are indications, however, that the summaries of the Pythagorean doctrines contained in the *Republic* and the *Metaphysics* may be in large part correct. The existent literature concerning Hippasus of Metapontum, the first disciple named in the tradition, suggest that he was both a mathematician and an important figure in the early field of harmonic theory.[5] The statements attributed to Philolaus of Croton, the oldest surviving Pythagorean texts, imply that he undertook various investigations of a similar sort, which involved both the study of numbers and the theory of musical proportions. One of his fragments maintains that "all knowable things contain numbers, without which nothing would be thinkable or knowable."[6] Philolaus may well have found evidence for this principle in the regularities of consonances, for a large portion of his texts consists of arithmetical analyses of intervals. "Harmony" (άρμονία) was for Philolaus, as for Empedocles and Heraclitus, the name of a principle of cosmic union, definable in the terms of natural relations.[7] But the early Pythagorean seems to have held these relations to be numerical in form. Moreover, as C. H. Kahn has written, "the distinctive feature of Philolaus' numbers is that they are arranged according to ratios that correspond to the three basic musical consonances," that is, the octave, the fifth, and the fourth.[8] Harmonic considerations seem to motivate the opening claim of his so-called "Fragment 1": "The nature of the cosmos results from harmony of the unlimited and the limited."[9]

Philosophers after Philolaus who wrote in the tradition of Pythagoras accorded like importance to the mathematical study of the laws of consonance. Archytas of Tarentum, a contemporary and alleged friend of Plato's, made of the affinities between the fields of arithmetic and music an explicit point of doctrine: the two investigations were, for him, "sister disciplines" (μαθήματα ἀδελφά).[10] That principle was to be recalled by the later thinkers who invoked Pythagoras. In the first centuries of the Common Era, when works claiming to record teachings of the old master of Samos began to proliferate, Pythagorean assertions of the unity of the study of number and sound became commonplace. In the second century A.D., when Nicomachus of Gerasa composed his influential handbooks of arithmetic and harmonics, there appears to have been nothing unusual about his decision to present them both as Pythagorean. By the time Iamblichus of Chalcis aimed to record the life and teachings of the archaic Greek thinker one century later, the doctrine had become canonical. Book 3 of his *Summa pythagorica* presents "music" (μουσική) as a discipline no less mathematical than arithmetic and geometry, the three being "rings in a chain that forms a single bond, as the most divine Plato says."[11]

It would be an error to infer from such facts that the ancient authors took Pythagoras to have been a "musician" in any contemporary sense of the word. Music appears to have interested the Sage not as an instrumental practice of art, but as one branch of mathematical knowledge, in itself both certain and necessary. Boethius's *Fundamentals of Arithmetic* is particularly illuminating in this regard. Its opening pages offer an account of the fields of study known to the Pythagoreans. We learn that "all the ancient authorities" in the tradition concur in understanding mathematical knowledge to be, in truth, the "wisdom" after which philosophers strive. Defined according to its nature, "wisdom" consists in "the comprehension of the truth of things that have their own, immutable substance," namely, "those things which neither grow by stretching nor diminish by crushing, nor are changed by variation, but are always in their proper force and keep themselves secure by support of their own nature."[12] Boethius names such entities in an extended list: "they are qualities, quantities, relations, acts, dispositions, places, times." All are "incorporeal," possessing an "immutable substance that cannot change." Admittedly, they may be "affected by the participation of a body and by contact with some variable thing." Then they pass, in particular beings, "into a condition of inconstant changeableness." But a quantity, like a quality, relation, act, or disposition, may also be considered in itself, as a pure object of thought. Such an ideality, we learn, is most properly called an "essence."

Boethius proceeds to explain that such essences are of two kinds. "One is continuous, joined together in its parts and not distributed in separate parts, as a tree, a stone, and all the bodies in this world." Such continuities, he explains, "are properly termed *magnitudes* [*magnitudines*]."[13] "The other essence is of itself disjoined and determined by its parts as though reduced to a single collective union, such as a flock, a populace, a chorus, a heap of things, things whose parts are terminated by their own extremities and are discrete from the extremity of some other. The proper name for these is a *multitude* [*multitudo*]."[14] A continuous essence may be divided without limitation. "A magnitude, beginning with a finite quantity," Boethius explains, "does not receive a new mode of being by division; its name includes the smallest sections of body."[15] The line, the circle, the pyramid, for example, can be infinitely reduced in size; despite every decrease in extension, they will nonetheless retain their identities. A discontinuous essence, by contrast, may be increased without limitation. "Every force of a multitude, progressing from one point," Boethius notes, "moves on to limitless increases of growth." The flock, the populace, the chorus, for example, may all be augmented in number; being collections of discrete elements, the addition of units cannot modify their essence.

To the opposition between the continuous and the discontinuous, the magnitude and the multitude, Boethius adds also a further distinction, which concerns the mode by which each essence may be grasped in thought. Each ideality may be understood according to one of two forms: either "by itself" (*per se*) or "by another" (*per aliud*). This difference can easily be explained. A continuous being, such as a line, a figure, or a solid, may be defined as a magnitude "in itself," for it bears no relation to another. But there also exist spheres in movement, which cannot be represented if not with respect to others, being "always turned in mobile change and at no time at rest." Similarly, certain multitudes may be said to exist purely "in themselves." Examples are "three, four … or whatever number which, as it is, lacks nothing." Such quantities, however, may also be conceived with respect to each other. Arithmetical ratios contain the proof. The identity of the "duplex" lies in the relation of two to one; that of the "sesquialter," in the relation of three to two; that of the "sesquitertial," in the relation of four to three.

Such an account of the nature of essences is hardly unique to Boethius. Similar classifications of the ideal objects of philosophy may be found in treatises of several ancient Neoplatonists, such as Nicomachus, Iamblichus, and Proclus. The theory of magnitudes and multitudes presented in the *Funamentals of Arithmetic* is remarkable above all for the disciplinary taxonomy that it introduces. Developing the ancient idea of a "cycle of study" (ἐγκύκλιος παιδεία), Boethius draws from the doctrine of essences the elements that allow him to offer a systematic account of the types of mathematical knowledge.[16] These types are four in number and, coining a term that was to have a long history, Boethius argues that they constitute a single "fourfold path" (*quadrivium*).[17] Each stands in relation to one "thing that has its own, immutable substance"; each is therefore "philosophical," or, more simply, mathematical. Magnitudes in themselves, we learn, are the proper objects of geometry. Magnitudes in relation to others, by contrast, belong to astronomy, which considers the continuous spherical beings that pass through the heavens. Multitudes contemplated in themselves are the ideal objects of arithmetic. Multitudes, finally, conceived with respect to each other belong to the domain of study that is "music" (*musica*).

For Boethius, music is thus a domain of mathematics, which considers essences no less ideal, necessary, and unchanging than those of arithmetic, geometry, and astronomy. More precisely, music constitutes a knowledge of multitudes, being in strict correlation with arithmetic. The implications of this fact are far-reaching and less evident than it might at first appear, if only on account of that deceptively familiar word, "arithmetic." For the moderns, the term designates an art of computation, a technique by which one employs numeric figures to calculate an exact measure. The ancients had an essentially different understanding of the word. Beginning with the Socratics, Greek mathematical thought distinguished between two forms of knowledge by which numbers (ἀριθμοί) may be apprehended: "arithmetic" (ἀριθμητική), on the one hand, and "logistic" (λογιστική), on the other. Rewritten in many ways, this distinction was to remain fundamental during the ancient period. "Logistic" was the name given to arts of measuring and calculating quantities of sensible things by means of numbers. "Arithmetic" was a knowledge of a different order. It concerned not the correct use of numbers but the true knowledge of their nature.

Several of Plato's dialogues already draw a contrast between the two domains. In a passage of the *Gorgias* reproduced in part in the *Charmides*, Socrates declares "arithmetic" to be the science that has as its object "the even and the odd, with reference to how much either happens to be"; "logistic," by contrast, "studies the even and the odd with respect to the multitude they make both with themselves and with one another."[18] That statement is difficult to interpret and has led to more than a single reading. But it suggests that arithmetic and logistic constitute the studies of two differing aspects of number: on the one hand, its properties as a pure quantity ("the even and the odd ... with reference to how much each happens to be"), and on the other, its properties in the composition of multitudes ("the even and the odd with respect to the multitude they make both with themselves and with one another").

The distinction between logistic and arithmetic was to grow at once simpler and sharper for the Neoplatonists, whose works

Boethius knew well. These thinkers contrasted numbers as the objects of pure thought (νοητά) with numbers as the objects of the senses (αἰσθητά). Arithmetic, they taught, studies the first; logistic handles the second. In his commentary on Euclid, Proclus thus distinguishes between skill in logistic and skill in arithmetic by explaining that "the man skilled in logistics does not study the properties of numbers as they are in themselves; rather, he studies them in the objects of sense."[19] An anonymous Neoplatonic scholium to the *Charmides* contains a similar teaching: "Logistic is a science that concerns itself with counted things, but not with numbers, not handling that number which is truly a number."[20] And Olympiodorus, in an influential scholium to the *Gorgias*, explains that "it must be understood that the following difference exists: arithmetic concerns itself with the kinds of numbers, logistic, on the other hand, with their material."[21]

When Boethius relates music to arithmetic, he thus draws the study of sound close to not calculation but a knowledge of the properties of numbers. Only questions involving the definition of the essence of multitudes, such as the characters of being even or odd, fall within the province of such an "arithmetic"; all considerations involving the uses and applications of numbers lie outside it. But there is more, for to grasp the mathematical nature of music as the Pythagoreans understood it, one fundamental ambiguity must still be dispelled. It is that of the term "number" itself. There are two good reasons to maintain that with the words *arithmoi* and *numeri*, the ancient Greek and Latin thinkers designated beings quite unlike our modern "numbers."

The first reason may be inferred from the definition of the antique *numeri* as "multitudes." This claim demands that every "number" be "disjoined and determined by its parts as though reduced to a single collective union," being similar to "a flock, a populace, a chorus, a heap of things, things whose parts are terminated by their own extremities and are discrete from the extremity of some other." In other words, every *numerus* must be discrete and discontinuous—in a word, it must be whole. Quantities not expressible as "natural numbers," therefore, cannot be *arithmoi*.

The second reason, too, follows from Boethius's terms, which are traditional. Discontinuous as a whole and in each of its elements, an *arithmos* constitutes a collection of units, being composed of many ones. That simple fact implies a startling consequence, whose importance can hardly be overestimated: namely, that the classical "number," *arithmos* or *numerus*, must always be more than one. Centuries before Boethius, Aristotle had already stressed this point: the smallest "number" of things, he explained, is two, the indivisible unit being smaller still.[22] At the inception of his book treating of *arithmoi*, Euclid himself had said no less: "a number is a multitude composed of units." [23] In ancient and medieval thought, *numeri* are therefore always sets of two or more elements of "one."

It might be objected that such an "arithmetic" is both ideal and abstract, for where in the physical world might one ever find a collection "disjoined and determined by its parts as though reduced to a single collective union," each as discrete and discontinuous, with respect to others, as the numbers on a line? We have been taught that continuity is the law of nature, which makes no leaps. That classical *numeri* are ideal, admittedly, can hardly be disputed; unchanging by definition, they in no way depend on physical bodies, properties, and occurrences, even if, to be perceived by us, they must be joined to matter. But for the classical thinkers, the eternity of such essences could hardly be said to render them unreal. Quite to the contrary: as a rule, "knowledge" in the sense of the ancient term *epistēmē* grasps only what is necessary and immutable. Moreover, it can be doubted whether the numbers of classical arithmetic, while ideal, may be considered to be "abstract," for unlike the signs familiar to modern mathematics, the ancient *numeri* are by definition beings. They are, to paraphrase Jacob Klein, "definite quantities of definite things," all reducible, as multitudes, to the indivisible essence that, more than any other thing, may be said to be: the "unit" or "monad" (μόνας), which is the form of "one."[24]

For the Pythagoreans, however, the reality of mathematics did not need to be demonstrated solely by argument. It could also be illustrated by experience. In the field of continuous quantities, astronomy furnished a proof: the stars in the heavens could be seen to move in accordance with the laws of magnitude. Physical phenomena might be invoked as illustration of geometrical regularities, if only high above this changing world and well beyond our sublunary realm.[25] Yet the decisive evidence for the reality of essences lay in not magnitudes but multitudes. Purely continuous quantities might well be viewed in the heavens; yet discontinuous quantities, Pythagoras showed, could be perceived on earth. The evidence, of course, lay in sound.[26] In the facts of music, in the relations of intervals, the essence of numbers could be detected; their properties and laws could be established. Pythagoras's findings in the forge had revealed this truth: the book of nature, he discovered, was written in the language of arithmetic.

For a good two thousand years, thinkers studied the pages of that book, refining their skills in reading and increasing their acquaintance with its letters. From the time of Philolaus and Archytas to the waning of the Middle Ages, harmonics remained the branch of philosophy in which it could be shown, to considerable effect, that sensible nature could be cognized by mathematical means. Boethius, at the midpoint of this long history, reflected the aims of the tradition. His *Fundamentals of Music* was both the fullest record of the ancient teachings and the single most important source for the medieval investigations into the art of consonant sounds. In large part because of this treatise, the Pythagorean doctrines far outlived antiquity.

Three basic consequences for the study of sonorous phenomena were to follow. Each may be derived from the fundamental correlation of music and arithmetic. Most obviously, harmonic sounds would be grasped in their discontinuity. Whether defined by pitch or by duration, acoustic phenomena would be studied to the degree to which they were, in essence, discrete in quantity. Thus they would also be transcribed, once a system of notation had developed. Pitch and duration would be considered to be composed of many counted units, similar in this to "a flock, a populace, a chorus... whose parts are terminated by their own extremities and are discrete from the

27

extremity of some other." Continuous phenomena, by contrast, would lie outside the domain of *musica*, which considered multitudes, not magnitudes, and related *numeri* to each other.

The second and third consequences are subtler but no less important. Just as in arithmetic, numbers were defined as the disjointed collections of unities, so in music, intervals would be treated as assemblies composed of many "ones." One unit of harmonics was easily identifiable. It was the single or "whole" tone. Aesthetic and acoustic evidence could be found for its structural autonomy. According to Nicomachus, Pythagoras, returning home from the forge, had already observed that the octave, the fifth, and the fourth are by nature pleasing. Moreover, he saw that every one of their combinations is in itself delightful— with a single, decisive exception: "what lies between the fourth and the fifth," Nicomachus noted, "is dissonant in itself."[27] That expanse is that of a whole tone. In a fragment preserved by Theon of Smyrna, Thrasyllus explained the oddity: "Notes are discordant...if the interval between them is that of a tone," he wrote, "for the tone is the source of concord, but not [itself a] concord."[28] Like the "monad" of arithmetic, the single tone of music was thus to be distinguished from the fourth, fifth, and octave, as the source of concord could be told apart from concord itself. The one and the many, the unit and the collection, were audibly distinct; sounds, once more, illustrated a principle in the doctrine of multitudes.

Implicit in this distinction lies the third and final consequence of the definition of music as a mathematical knowledge intimately linked to arithmetic. To remain faithful to Pythagoras, the classical and medieval theorists of harmony dared not venture beyond the limit that defined the old metaphysics of numbers. For the theorists of *musica* and *arithmetica*, it remained an axiom that collections be assemblies of something—more precisely, that they be assemblies of "ones." Whether considering absolute multitudes or multitudes in relation to others, whether contemplating numbers in themselves or numbers perceptible in sound, the Pythagoreans always counted one by one, looking toward the ideal of unity: the "least element of quantity,"

28

as Iamblichus taught, "the primary and common part of quantity, or the source of quantity."[29] In deference to that "source," the classical and medieval thinkers refused to concede that, in the domain of music or arithmetic, there could be things irreducible to unity, beings unequal to one. As a matter of principle, they would not grant that the art of sounds admitted musical beings irreducible to the laws of number.

The Pythagorean investigations, however, cast a shadow over that belief. More than once, and almost despite themselves, the disciplines of Pythagoras encountered sonorous phenomena that eluded the grasp of disjunctive essences and that no arithmetical relation—no matter how complex—might fully define. The shock of a conflict in the ordered world, then, could hardly be avoided: the ancient *arithmoi* ran up against a nature that was real, yet uncountable by them. Wherever possible, the Pythagoreans resolved not to name that immeasurable reality. Refining the tools of their art, they developed means to moderate it as best they could, reducing asymmetries to the ordered inequalities of measured multitudes. Despite the Pythagoreans' best efforts, however, their project was not to succeed. Ratios might be sought and found beyond all audition and all imagination. Yet something numberless continued, obstinately, still to sound.

CHAPTER THREE

Remainders

For the ancient Greeks, the most minimal of musical accords was the fourth. It was the smallest consonance said to have been heard by Pythagoras, and it was the basis of the classical scales, which grew from its ordered tones. According to Aristoxenus, the fourth-century thinker whose *Elements of Harmony* contains the most detailed surviving account of the subject, the music of the ancient Greeks was conceived in an intervallic expanse of two octaves. Each octave was composed of two fourths, or tetrachords, joined by one tone.[1] Within the fourth, two outer positions were fixed, while two inner ones varied. "Attunements" (ἁρμονίαι) would result from the positions in the tetrachord. For example, one temperament was defined by containing a tone, another tone, and a third interval, smaller than that of the tone; this was the "diatonic" kind. Another tuning consisted of two even smaller intervals, followed by an interval greater than that of a single tone; the ancients called this the "enharmonic" genus. Finally, if the tones were arranged so as to form an order containing one interval greater than that of a tone, followed by two pitches separated by less than a tone, the "chromatic" attunement would be produced. These were the three most famous arrangements of sounds within the musical expanse of the tetrachord. Yet they were far from being the only orderings of sounds known to the Greeks. The intervals within the fourth could be further varied in their structure. One set of four could be appended, through the addition of a tone, to a second set of four, different in kind from the first; similarly, an entire octave, with its succession of pitches, could be attached to a second octave, tuned according to other harmonic shapes. In every case, however, considerations of harmonic order departed from the elementary "system" (σύστημα) that was the tetrachord. Theorists of music invented for it a name to indicate that it constituted the unit of musical sound. Long before the rise of the art of grammar, Philolaus and those who followed him named the fourth "what is taken together," or, more simply put, the "syllable" (συλλαβή).[2]

The simplicity of that consonance revealed itself, however, to be deceptive. From the difference of the fourth with respect to the fifth, the ancients had deduced the unit of harmonic construction that was the single tone. One might well expect that some number of this basic interval would compose the fourth. But how? Two tones were clearly too little to add up to the desired consonance; three, by contrast, were just as evidently too many to form a single "syllable." Musicians familiar with the ways of instruments could draw this simple conclusion: no more and no less than two and a half tones compose the interval. They could argue, in other words, that the space between a C and an F coincides with that of C to D (one tone), D to E (one tone), and, finally, E to F (a half tone). Such a series would be "diatonic"; the total interval would be that of the fourth. Aristoxenus advanced this thesis explicitly in his *Elements of Harmony*: "The tone," he wrote, "is that by which the fifth is greater than the fourth; the fourth is two and a half tones."[3] Today, the statement may seem evident; the modern major scale begins with precisely such a diatonic progression, and it suffices to observe a keyboard to surmise that every tone may be divided into two. But Aristoxenus's argument about the fourth implied one major proposition that, to the ancients, was anything but self-evident. His claim clearly suggested that to define the extension proper to the fourth, one must imagine parts of the single tone, admitting, simply, that this basic interval may be halved.

The Pythagoreans could accept no such proposition. It might be recalled that in the forge, the master had perceived three consonances and one dissonance, from which four intervals could be derived. The smallest of them all was the single tone. A "half tone" had not been recorded. The reason for this fact was far from incidental. To partition a tone, it is necessary to represent it as a quantity susceptible to such division. Ancient thinkers, of course, were well acquainted with things whose quantities might be continually reduced at will. They called them "magnitudes." Geometrical beings such as the line, the circle, and the cube were, for them, of this nature. They possess the quality of being such that, no matter their quantity, they may always be represented as smaller. No line, figure, or solid, after all, alters its nature in being diminished in size. There is evidence that some classical thinkers held sounds also to be of this kind. The "harmonists" (ἁρμονικοί) mentioned in several sources may have been among them.[4] Proposing the halving of the tone, Aristoxenus may also have shared that view.[5]

Yet the followers of Pythagoras could in no way concur, and for a simple reason: they held musical phenomena to be not magnitudes, but multitudes. Like all things arithmetical, harmonic intervals, for the Pythagoreans, could always be increased in quantity: multiplication contained the proof. But it could by no means be assumed that intervals could be reduced *ad infinitum*. Pythagoras had taught that the reasons of the consonances were to be sought in the ratios of *arithmoi*. From this principle, it followed that in reducing such ratios by division, one would eventually run up against a lower limit in which one number stood in relation to another in the most elementary of possible forms. From "twelve to six" (12:6) for example, one might well produce "twenty-four to twelve" (12:24), and "twenty-four to forty-eight" (48:24) and so on, in successive doublings; all such relations represent the basic interval of the octave. By division, one might also pass down to "two to one" (2:1), a relation that expresses the same basic inequality. Yet having reached that fundamental form, one could go no further without abandoning the form of the ratio as such.

Like Aristoxenus long after him, Pythagoras had been known to profess the doctrine that "the tone is that by which the fifth is greater than the fourth." But for the archaic Greek thinker, that statement implied a precise mathematical proposition: the tone, he showed, consists in the exact multitude that remains when one removes a fourth from a fifth. One technical point is worth recalling: because the classical musical quantities were represented as intervals, to "subtract" them, one must divide them; to "add" them, one must multiply them.[6] Given these rudiments of "logistics," it can easily be shown that when one "withdraws" a fourth (4:3) from a fifth (3:2), one tone remains, defined by a single relation: the "sesquioctave," or the arithmetical ratio of nine to eight (9:8). For the classical thinkers, this relation of quantities belongs to a particular class of arithmetical "inequalities," or ratios said to be "epimoric" or "superparticular." Such relations are defined by the fact that, as Boethius would later explain, "the larger number contains the smaller number, in addition to one part of its parts."[7] In more formal terms, one may therefore also define these relations as possessing the mathematical shape of $n + 1:n$. Each of the three intervals that Pythagoras identified in the mythic forge shares this arithmetical nature: the octave (2:1), the fifth (3:2), the fourth (4:3) and the tone (9:8) can all be represented by "superparticular" or "multiple" relations.[8]

It is hardly surprising that the Pythagoreans devoted particular attention to this class of arithmetical inequalities. In the fifth century, Archytas of Tarentum developed a mathematical theorem that bore on precisely such relations. Boethius recorded it in the *Fundamentals of Music*. But it also appears, more succinctly, as a basic proposition in the Euclidean treatise known to the tradition as *The Division of the Canon* (*Sectio canonis*). The Pythagorean demonstration led to a single, unequivocal conclusion: numbers in "superparticular" relationship cannot be divided by any geometric mean. In other words, for ratios of two integers, in which the larger integer exceeds the smaller integer by one of the smaller integer's parts, there is no third integer that stands in an equal relation to both.[9] For the numbers one and nine, by contrast, such a third integer can be found: it is the number three, said to be their "mean proportional." Archytas demonstrated that none of the basic harmonic intervals admits of such a mean. As his theorem reads in its Euclidean form,

"Neither one number, nor many numbers, can be interpolated in continued proportion in an epimoric interval."[10] This rule holds for the interval of the tone by virtue of its arithmetical structure. In the *Fundamentals of Music*, Boethius explained the law in terms similar to Euclid: on account of its essence, the ratio of nine to eight (9:8), being "superparticular," "cannot be divided into equal parts by the interpolation of a mean proportional number."[11] Thus the "whole" tone, Archytas proved, could not be halved.

No doubt the Pythagoreans met with resistance of an empirical variety. One imagines that a theorist might claim to perceive two equal intervals within the expanse of a single tone; a musician might divide a tone in two, or even four. As Aristoxenus presents them, the Greek tetrachords, or "fourths," demanded no less. But it is difficult to imagine that such rebuttals could persuade the philosophers and mathematicians who had followed Archytas in the demonstration of his theorem. The tone could be arithmetically represented by the proportion of nine to eight (9:8); that relation could be shown to be "superparticular" in kind; it could be proved, finally, that such inequality cannot be divided by any mean proportional. The doctrine of the indivisibility of the tone was thus not a matter of opinion; it was the ineluctable consequence of a proof. Yet one problem remained unsolved. The interval of the fourth was clearly greater than that of two tones; it was also undeniably much less than three. But how much, then, might it be said to be? Aristoxenus offered a clear answer to the question in claiming that the tetrachord was made of "two and a half tones." The Pythagoreans pronounced that solution untrue, as well as arithmetically incoherent. Yet they thereby obligated themselves to find another answer to offer in its place.

Faithful to the principle of number, the Pythagoreans resolved the quandary by availing themselves of their logistical craft. To find the exact quantities of tones that compose the fourth, it sufficed for them to "withdraw" from the fourth two tones. Originally, that operation may have been difficult to perform, but by the fifth century, if not sooner, it had been done.[12] Then, it could be shown that, after the "subtraction" of two tones from the tetrachord, an interval remained that was less than half a tone. This was an inequality that could be defined by a complex arithmetical ratio: two hundred and fifty-six with respect to two hundred and forty-three (256:243). That seemingly arcane relation was to be found not only in works on harmonics. Plato's famous investigation into cosmology, the *Timaeus*, contained it too. Explaining the secrets of the world soul, Timaeus recounts that in his act of making, the demiurge chose to "fill up" the interval of the fourth (4:3) with intervals of the tone (9:8). An "interval," he explains, then "remained," and it was equal to two hundred and fifty-six to two hundred and forty-three (256:243).[13] In deference to that passage, thinkers after Plato called this subtle ratio *leimma*, or "remainder" (τὸ λεῖμμα).[14]

In the analysis of the harmonic structure of the fourth, however, this was not the only such multitude to be a "remainder." The Pythagoreans went further in their arithmetical division of musical sound. The discrete quantity of the *leimma* could be compared with that of the tone; the difference between the two could consequently be measured in a ratio. They called this inequality the *apotomē*, literally, "what is cut off." At the limits of imaginable relations, this interval could be expressed by the ratio of exactly two hundred and seventy three and three eighths to two hundred and fifty-six (273³/₈:256, or 2187:2048). Venturing still further into the subtleties of logistics, the Pythagoreans also envisaged the difference between the *leimma* and the *apotomē*. They called this last remainder "the comma" (τὸ κόμμα), "what is struck out," defining its ratio as 531,441:524,288.[15]

It has been noted more than once that such ratios are difficult to grasp, and some modern commentators have not hesitated to dub them fantastical. Citing the arithmetical relation that defines the comma, Walter Burkert, an unrivalled authority on the Pythagoreans, could not keep himself from remarking: "pure frivolity!"[16] Such judgments echo those of several ancient authors. Aristoxenus refused to measure any interval less than the "semitone" (δίεσις) on the grounds that "the voice cannot produce it, nor can the hearing

36

detect it."[17] Doubtless, the Pythagorean divisions of the tetrachord were not intended for use in the arts of instrumental practice or in observation. A string player, for instance, might well halve a string, sounding the relation of two to one and producing the interval of a single octave. Anyone could hear it. But it is difficult to imagine a musician executing the arithmetical ratios that define any of the three slight intervals in the fourth, whether *leimma*, *apotomē*, or *comma*. Moreover, it is impossible to conceive of their being perceived as such by any human ear. It is not that the ancient Pythagorean inequalities are by modern acoustical standards lacking in exactitude; on the contrary, they are quite precise. Yet the Pythagorean precision is not for us. In excess of human sensation, it points to truths that only a calculus of quantities can reveal.

This fact renders the ratios of measured multitudes less "frivolous" than they might seem. In their arithmetical excesses, the relations indicate how far the transliteration of nature by classical numbers could lead. Pythagoras may have begun with the distinct perception of arithmetical ratios in sounding bodies. Once his disciples committed themselves to the mathematical study of nature, however, they found that, at a certain point, the ways of conscious perception and arithmetical consideration must part. Then, to the perplexity of their contemporaries and later commentators, the disciples of Pythagoras resolved to follow an unexpected path: they renounced the evidence of their senses for the certainty of their arithmetic.

The precision of ancient numbers, however, could not be limitless. *Arithmoi* and *numeri* signified equalities and inequalities of varying complexity, from the simple to the subtle. But as the signs of multitudes in relationships, such figures always pointed to beings that were essentially discrete and whole. "Nine to eight," "four to three," and "three to two" were for the ancients not abbreviated expressions for arithmetical quantities lying between integers on a number line, such as our modern rational numbers. "Three to two," for instance, was for them not equal to the number "1.5," for the Greeks and Romans knew no such "number." The classical inequalities, instead, signified ideal and eternal arrangements, in which two sets of discrete units stood in relation to each other. For this reason, it would be an error to believe that such ratios as "three to two" and "four to three" denoted "fractions" in the modern sense of the term. As their name indicates, such entities represent the fracture of the principle of unity into a multitude of parts, something classical arithmetic, in principle, would not admit.

Sometime close to the end of the first century A.D., Theon of Smyrna explained this point: for a theory in which numbers are defined as multitudes, divisions may be referred to counted things, but not to the unit of counting itself. In his treatise on *Mathematical Knowledge Useful for the Reading of Plato*, Theon explains that nothing is simpler than to envisage the division of things into parts. But the division of *arithmoi* is another matter. While one may break every number into an assembly of separate "ones," one cannot divide the monad itself, a "part of one" being a sheer contradiction in terms. To illustrate his thesis, Theon suggests that one consider the breaking of a sensible thing into parts. When it is divided into multiple portions, a thing passes from being unitary to being multiple, in accordance with its matter, which renders it divisible to infinity. But the arithmetical unit itself—"one"—remains essentially undivided. Not only is its unity not broken; as Theon notes, it is multiplied, for where there was once a single identifiable unit, there are, after division, a number of "ones," which form an assembly of separate parts. "The one," Theon concluded, can be therefore diminished only "as a body"; "as a number," it can solely be augmented.[18]

The ancient harmonic inequalities are best understood in such terms. As an example, one may take the Pythagorean "remainder" defined by the ratio of two hundred and fifty-six to two hundred and forty-three (256:243). Were one to fix the arithmetical value of this interval by modern means, one might divide the two quantities; from that operation one would obtain an exact, if cumbersome, rational number slightly greater than 1 (1.053497942386831). Were one to determine the value of this interval in terms of the acoustical system most often employed in musical tuning today, one would convert the old ratio into the units of sound E. J. Ellis invented in the nineteenth century; by means of a logarithmic calculation, one would conclude that the *leimma* possesses the acoustical size of 90 cents, where the whole octave contains 1200 such units.[19]

Today, such translations of the ancient notations are common, since they allow the old musical intervals to be compared, at a glance, with the modern. Yet to represent the classical ratios solely in such a form is to overlook the reasons for their conception and the meaning of their definition. For the ancients, the notation "256:243" did not point to any quantity "more than one and less than two," nor did it signify any being, such as those measured in geometry, that is susceptible to continuous increase and reduction. The old notation represented the relation of two ideal multitudes, made of ones, which, when found in the proportion of bodies such as strings, produced an audible consonance. If the transliteration of intervals by arithmetical inequalities persisted for centuries in antiquity and the Middle Ages, it was because this system allowed for an analysis of harmonies that reckoned solely in integers and their relations.

The writing of ratios, therefore, was no detail in the Pythagorean arts. Relations entered arithmetic and music precisely where "numbers" could not be. Setting one discontinuous collection of units against another, they allowed the ancient thinkers to designate quantities that would otherwise have remained ungraspable as such, being solely expressible by the relations of multitudes. Yet the ancient thinkers might avail themselves of such expedients only to a point. Quantitative relations could be proposed for some intervallic divisions, such as in the decomposition of the diatonic tetrachord into two tones and one *leimma*. Other intervals resisted such division. In this sense, the debated halving of the tone was less exceptional than exemplary. Since all the ancient consonances were "superparticular" or "multiple" in arithmetical structure, it followed from Archytas's theorem that neither one number nor many numbers could be interpolated in continued proportion within any of one of the three Pythagorean accords. This meant that no basic interval could be represented in equal partition by a ratio; the octave, fifth, and fourth, in other words, could not be "halved" by any number of ancient numbers, even when placed in relation to each other. Were the exact midpoints of such accords in any way to exist, they would be, therefore, beyond music and arithmetic alike. Neither consonances nor tones, neither intervals nor their elements, they would be not only imperceptible, like the *leimma*, *apotomē*, and *comma*. They would also be uncountable and, therefore, rigorously indefinable.

The Pythagoreans, however, were no strangers to the uncountable. Although they barred numberless relations from the domain of their arithmetic, they also named them in no uncertain terms. They called them "unspeakable" (ἄρρητοι), "irrational" (ἄλογοι), and "incommensurable" (ἀσύμμετροι). From such appellations, one might infer close acquaintanceship. Yet the familiarity the classical theorists of number possessed with such relations could not be knowledge, according to any classical standards of science. Infinitely eluding the rule of unity, incommensurable quantities could not be considered to number anything that was and that remained a single thing; for this reason, they could hardly be considered to number anything at all. Of such unspeakable relations, it could only be deduced that, like the impossible root of the single tone, they could be no collections of one. They were, quite simply, immeasurable, and as long as every definition in arithmetic and music was to be numerical and every number was to be discrete, they were unrepresentable as such. They might well have been somehow manifest to the Pythagoreans, but, being uncountable, they could be no "remainders." Their sole place was at the limits of their art of quantity. For them to be shaken from those outer edges, for them to come to be considered in themselves, the definitions of music and arithmetic, with their aims and methods, would need to be recast in their basic terms. Things once "unspeakable" would need to be spoken. These were necessities that the early Pythagoreans would hardly have granted. In time, however, such demands made themselves felt with ever greater force, until new elements of quantity were found and, by means of them, thinkers came to notate tunings unheard in the ancient and the early medieval world.

CHAPTER FOUR

Disproportions

According to several traditions, it is the Pythagoreans who discovered that within the world of nature, there exist both proportions and disproportions, and that between commensurable and incommensurable relations, the difference is absolute. Yet numerous records also indicate that the Pythagoreans never intended to divulge their knowledge of such matters to those uninitiated into their teachings. It is said that a Pythagorean revealed to the ancient world the existence of mathematical "irrationalities," but it is also said that he was punished as a result. Introducing the doctrine of the nature of commensurability and incommensurability, Iamblichus recounts that the followers of Pythagoras "profoundly hated the man who first revealed the teaching to those who were not worthy of sharing their doctrines. Not only did they banish him from their community of study and their way of life. In addition, they built for him a tomb, as if the one who had once been their companion had truly departed from the life of men."[1] Other reports suggest that as a consequence of his impious deeds, the Pythagorean suffered a far more violent fate: the gods drowned him at sea, as they did the iniquitous disciple who showed how to construct a dodecahedron.[2] Close to the late third century A.D., Pappus of Alexandria interpreted these reports in his commentary on book 10 of Euclid's *Elements*, which today survives solely in Arabic. The knowledge of incommensurables, Pappus explains,

> had its origin in the school of Pythagoras, though it underwent an important development at the hands of the Athenian, Theaetetus....

Indeed, the school of Pythagoras was so affected by its reverence for these things that a saying became current in it, namely, that he who first discovered the knowledge of incommensurables or irrationals and spread it abroad among the common herd perished by drowning. This is most probably a parable, by which they sought to express their convictions: first, it is better to conceal every incommensurable, or irrational, or inconceivable in the world and, second, the soul that, by error or heedlessness, discovers or reveals anything of this nature that is in it or in this world, wanders thereafter to and fro in the sea of non-identity, immersed in the stream of becoming and decay, where there is no standard of measurement.[3]

One may imagine why the ancient Pythagoreans may have considered "the knowledge of the incommensurable" best concealed. That doctrine involved the limits of the theory of unities. Once it was disclosed that certain quantities are incommensurable among themselves, it would be seen that there are mathematical ratios that are strictly indefinable by integers, as well as by all their "inequalities." It would be known that numbers cannot transcribe the measures of this world. This may have been the truth the thinkers of *arithmoi* wished to keep a secret. Yet it remains less than evident how and when the ancient Pythagoreans first came upon their secret knowledge.

A number of hypotheses have been proposed.[4] Many scholars have argued that the incommensurable was discovered in geometry. The procedure of "anthyphairesis," also known as "reciprocal subtraction" or "the Euclidean division algorithm," may have played a role in the revelation of the arithmetical imbalance. Commenting on Euclid's propositions, W. R. Knorr has summarized the classical geometrical procedure in these terms:

> Two homogenous magnitudes *A* and *B* are given. The smaller (say, *B*), is subtracted from the larger, leaving the remainder *C*. If *C* is smaller than *B*, it is subtracted from *B* to produce a second remainder. If *C* is larger than *B*, then *B* is subtracted from *C*. In either case, a new remainder *D* is obtained, and it is used in the same fashion with respect to the previous subtrahend, yielding a new remainder.[5]

If this operation is applied to lengths measured by whole numbers, it must end in a finite set of steps. The last geometrical "remainder" that may be measured by an integer will be the smallest common divisor of A and B. So, too, if the procedure is applied to commensurate magnitudes, it will terminate, ending with a remainder that is the least common measure of A and B. But there are also cases in which the procedure of reciprocal subtraction will continue without end. Infinite in operations, the method will then be algorithmic. By successive decrease, the remainders will grow shorter than any assigned finite magnitude, yet no portion will have exhausted the measured segment. Such is the case of two magnitudes in "mean and extreme ratio," or, to use a more familiar expression, two lengths divided by a "golden" mean or section, such that the smaller is to the greater as the greater is to the smaller and greater together.[6] Between the shorter and the longer of two such magnitudes, no least common measure will ever be found. They are "incommensurable" in the precise sense defined by Euclid in Book 10 of the *Elements*, which treats of such quantities: "If, when the less of two unequal magnitudes is continually subtracted in turn from the greater, that which is left never measures the one before it, the magnitudes will be incommensurable."[7]

Other historians of Greek mathematics have traced the discovery of irrational magnitudes to the study of a single and far simpler geometrical figure. It suffices to contemplate the properties of squares to note a strictly incommensurable relation between two distinct continuous quantities. One may recall the theorem now called "Pythagorean," which was known to the early Greeks without, it seems, ever having been attributed to Pythagoras himself.[8] This rule dictates that for any right-angle triangle, the square length of the hypotenuse is equivalent to the sum of the square lengths of the two sides (in other words, for a right-angle triangle whose sides are denoted by A, B, and C, where C denotes the hypotenuse, $C^2 = A^2 + B^2$). A square divides into two such right triangles; the relation of the triangle's sides to its hypotenuse is equivalent to that of the square's sides to its diagonal. From this elementary principle, it follows that if the side measures one, the diagonal will measure that quantity whose square is two. Yet such a quantity, for the ancients, could be no "number." Infinitely irreducible to unity, it cannot be expressed by the ratio of two integers. The asymmetry between side and diagonal is both apparent and absolute, and Plato and Aristotle named it often when discussing incommensurability. In different ways, the *Theaetetus* and the *Meno* both find Socrates investigating the implications of the relation between the side and the diagonal of the square.[9] And in the *Prior Analytics*, Aristotle presents a theorem for incommensurability that elegantly demonstrates, in a few simple steps, that if one takes the diagonal to be commensurate with the side, arithmetical absurdities must follow. By means of no more mathematical knowledge than "the Pythagorean theorem," the diagonal can then be shown to be both even and odd, thus proving its arithmetical incommensurability with respect to the side.[10]

Aristotle's is the oldest of the ancient theorems of incommensurability. Although it finds its clearest form in an apocryphal addition to Book 10 of Euclid's *Elements*, it may date back as far as to the fifth century B.C., since it is also attributed to the early Pythagorean mathematician Hippocrates of Chios.[11] It is remarkable that although it bears on a relation in a geometrical figure, the proof defines incommensurability in terms of the two great arithmetical notions: the "even" and the "odd."[12] That fact indicates that the discovery of the irrational may have been more related to arithmetic than the Euclidean algorithm suggests. There can be no doubt that the classical thinkers, in any case, hardly needed to leave the field of multitudes to come upon quantities inexpressible by numbers. The ancients could also have encountered such irrationalities in their investigations into the various means that can be interpolated between numbers. As Simone Weil once observed, it would have sufficed for the Greek thinkers to note that between one *arithmos* and its double, there cannot be any "mean proportional": then, the ancients would have run up against the "irrational."[13] The point merits serious consideration, but one may also build upon it and take one further step. The classical thinkers treated the study of numbers in relation to each other in the field of knowledge that was "music." There, the ratio of a number to its double bore a technical name. Represented in its simplest form as the relation of two to one (2:1), this interval was called "diapason": the ratio that, when referred to the length of strings, produces the octave. The ancients may well have come upon irrational magnitudes in their geometry, and they may equally well have first encountered such quantities in their arithmetic. Yet as Paul Tannery long ago noted, it is also possible that the early Greek thinkers discovered incommensurability in studying the art of harmony,[14] for in the moment the ancients sought to "divide" the octave, inserting a mean proportional within the expanse of the diapason, they would have discovered the very irrationality exhibited in the diagonal of the square. They would, then, have met the inexpressible root of the number two.

Whatever the conditions in which they came to light, however, incommensurable quantities were soon to acquire a stable place in the ancient mathematical arts. Geometry would be the field in which they could be studied—not as such, to be sure, for the "irrational" by definition could not be reasoned, but in proportions or "analogies," to which numbers did not need to correspond. After Theodorus and Theaetetus, Eudoxus of Knidos, Plato's pupil, developed the terms and theorems that lay at the basis of this field of study. With his "doctrine of proportions," Eudoxus showed that incommensurable quantities could be treated by a nonarithmetical mathematical discipline that would bear on all homogeneous magnitudes, whether lines, figures, or solids. Between one incommensurable magnitude and another, a single ratio still remained arithmetically indefinable, but now, nonetheless, it became possible to define one such incommensurable magnitude as mathematically equivalent to another. Thus, to take but one example, in a circle, the exact relation of the circumference to the diameter can be established only by means of an irrational magnitude: the quantity called π, which no ancient thinker would have considered a "number." Yet one may still establish geometrically that such a relation holds for all circles. Between two such magnitudes, one may therefore establish a "ratio

of ratios," or "analogy."[15] Once this principle had been accepted, the
threat of the irrational was in part contained: the incommensurable
could be defined as an object of geometry, and arithmetic could pro-
ceed undaunted in its consideration of multitudes, all commensurate
by virtue of their definition as diverse collections of many ones.
Boethius's *quadrivium*, divided between the arts of discontinuity and
the arts of continuity, number and size, multitude and magnitude,
rested upon this division.

When, after Boethius, the art of music developed in new ways
in the Middle Ages, it was to remain, even in its most astonishing
inventions, faithful to this partition. The history of notation may
offer the most striking illustration of this fact. The period from
the first half of the ninth century to the first half of the eleventh
century witnessed the emergence and development of a new system
of writing that permitted, with ever greater subtlety, the registra-
tion of liturgical chant. This new means of representation allowed
an oral repertoire to be preserved. Yet it was also to be an essential
element in the constitution of a different art. Thanks to the new
artifices of writing, it was soon possible to fix the positions of
several voices, which could be at once coordinated and yet in part
autonomous. Western "polyphony" would thus, in time, emerge. As
Marie-Elisabeth Duchez has shown, the notation that accompanied it
was no simple mirror of its reality. The transcriptions illustrated the
principles of not sound but its conception. From Pseudo-Alcuin and
Remi of Auxerre to Hucbald, the author of the *Musica enchiriadis*, and
Guido of Arezzo, the empirical reality of song was gradually "objec-
tified" and "conceptually abstracted" by a graphic practice founded
on five fundamental notions: "the perceptual and, later, quantitative concept of the interval (the tone)"; "the consequent notion of
discrete sound"; the idea of "the height of sound," which permitted
the classical distinction between the acuity and gravity of sounds to
be spatially exhibited on the vertical axis of the page; "the concept
of the *note*, resulting from the [...] preceding conceptualizations";
and, finally, "the notion of musical scale," which emerged from the
currents that gave rise to the four preceding concepts, as well from

46

two traditions of diagrams: those of ancient Greek music, as related by Boethius, and those of the Neoplatonic and Pythagorean doctrines of the harmony of the celestial spheres.[16] These five concepts all rest on the principle that to be musically intelligible, sounds must be essentially discrete in quantity, like the old multitudes of arithmetic. "The knowledge inherited from the Greeks," Duchez concludes, "in some way 'suggested' to the medieval musicians what they should be hearing in their chant, so that they might write it down. But medieval thought succeeded where Greek thought failed, establishing in music a rational and audio-visual link between mathematical theory and an evocative, graphic semiotics."[17]

That "link" was to bind more than the melodic aspects of medieval music. It also extended to rhythmic structure. Developing their "graphic semiotics," medieval musicians considered durations, as well as intervals, to be multitudes. This vision was to play a crucial role in the development of Western polyphony. "In Gregorian chant, at least in such a relatively late period as the 13th century," Willi Apel has argued, "the melodies were sung in more or less equal note values, without any concern about rhythmic differentiation."[18] Yet in polyphonic composition, a system of strict rhythmic orderings had developed by 1200, and it was soon to be a basic element in musical practice. Durations came to be represented as homogenous numeric quantities, such that musicians could contrast "long" tones to "brief," as three or two units to one. Conjunctions of *longae* and *breves*, in turn, gave rise to a set of rhythmic patterns. By the ninth century, if not sooner, Frankish cantors had abstracted from the repertory of Gregorian chant a system of melodic sequences. They thus created eight harmonic modes.[19] Once a stable notation for durations had emerged, it became possible also to establish rhythmic modes, composed of a fixed series of "longs" and "briefs." Those sequences could then be represented in various forms in musical works. In the theory of rhythmic notation developed by John of Garland, for example, repetitions of modal sequences would be classified into a series of "orders," each of which would be defined as the multiple of the first.[20]

47

The medieval authors went far in such numeric representations of time. In addition to measuring the duration of melodies in homogenous, divisible elements, they also developed the art of altering sequences of time through forms of augmentation and diminution defined in the classical art of arithmetic. In one motet, for example, a single rhythmic pattern might appear several times, in durations whose varieties reflected the ratios studied in ancient mathematics. In the early thirteenth century, Perotinus showed how the modal elements of a single melodic sequence might appear twice, first in one duration, then in an equivalent to its half or double.[21] By the second half of the fourteenth century, systems of writing emerged that allowed composers to modify temporal values with far greater subtlety. Special symbols could be employed to designate proportional increases and reductions in time without its being necessary to alter any of the notes.[22] Fifteenth-century theorists and composers, developing such signs, proposed a graphic system to indicate the proportional increase and reduction of durations. In their new notation, they could mark the points at which a sequence was to be increased or reduced by ratios of varying complexity, from 2:1 to 5:4 and 21:5. "In this period," Apel comments, "the system of proportions developed far beyond the bounds of practical application into the realm of pure speculation. [Franchinus] Gaffurius, for instance, does not hesitate to explain proportions calling for a diminution of 9:23."[23]

Such proportional "extravagances" are, of course, the hallmark of traditional Pythagorean arithmetical harmonics, but their application to the field of rhythm constituted an essentially medieval invention. The *Fundamentals of Music* offered the theoretical foundation for such procedures. Just as Boethius defined five types of ratio corresponding to five kinds of related multitudes, so in the fifteenth century, Guilelmus Monachus, Gaffurius, and Johannes Tinctoris would admit in their doctrine five rhythmic "proportions."[24] And just as, for the ancient thinker, multitudes were collections of ones, so for the medieval composers, such ratios of rhythms would be commensurable by nature. "The system of mensural notation," Apel writes, "rests upon the principle of a fixed, i.e., unchangeable unit of time, the *tactus*, a beat in moderately slow speed...which pervades the music of this period like a uniform pulse."[25]

In the same epoch, however, a major shift in perspective was underway. Harmonies were to be perceived not in numbers but in figures, and they would be measured not by multitudes but by magnitudes. Slowly, if irreversibly, the arithmetical foundations of the theory of proportional sound began to give way. The work of the fourteenth-century thinker Nicole Oresme is singularly prophetic in this respect. In his *Treatise on the Configurations of Qualities and Motions*, thought to have been composed close to the middle of the century, Oresme proposes a method for the mathematical study of sensible phenomena.[26] More precisely, he seeks to show how it is possible to quantify continuous beings, such as qualities, speeds, and motions, by strictly geometrical means. Oresme introduces his considerations in the most traditional of terms. "With the exception of numbers," he writes, "every measurable thing is imagined in the manner of continuous quantity. Therefore, for the measurement of such a thing, it is necessary that points, lines, and surfaces, or their properties, be imagined."[27] That thesis is in appearance classical, resting upon the ancient distinction between the two varieties of quantity: the discontinuous and the continuous, which may be attributed, respectively, to the fields of arithmetic and geometry. The remark, however, serves to introduce a novel development. Oresme's entire treatise aims to demonstrate that it is possible to "imagine" continuous beings thanks to a geometrical notation first invented in the Middle Ages.

For this Scholastic thinker, magnitudes, such as qualities and motions, possess degrees; in any one time or space, they imply a determinate "intensity." Moreover, being continuous by nature, intensities are susceptible to infinite increase and reduction, in time as in space. One may take as an example the quality of coldness. A body may grow cooler in some of its present parts than in others, and a single part may, in successive instants, become increasingly cold. Employing a conceptual vocabulary common to several fourteenth-century thinkers, Oresme argues also that such intensive variations occur in two ways. They may come to pass *uniformiter*; then, for each period of time that elapses or for each part considered, the reduction of temperature will remain constant. Conversely, they may occur *difformiter*; in such cases, the degree of change will alter in each part or moment studied. Moreover, Oresme specifies that a fundamentally "uniform" intensive change may be secondarily characterized as either regular or irregular, just as a basically "difform" alteration may be qualified, in a supplemental sense, as constant or inconstant. For all such occurrences, Oresme proposes a system of geometrical illustration, based in part on a model proposed before him by the philosopher Roger Bacon.[28] The method can be simply summarized. For each qualitative variation, two axes will be drawn. The first is horizontal. It illustrates the extension of space or time in which the alteration comes to pass. To this magnitude, which is divided into units termed "degrees," Oresme gives the name of "longitude" (*longitudo*). Perpendicular to any point along this segment, one may then assign a value of intensity, measured not in length, but in height, on a vertical axis, also in degrees. The absence of the quality under consideration will be represented at the same level as the horizontal line, and each successive augmentation will be marked by a point of increasing elevation. If one links the various points of intensive fluctuation over time or space, one will thereby obtain a second line. To this second magnitude, Oresme gives the name of "latitude" (*latitudo*).[29]

The method can be applied most easily to measure the variations of qualities in surfaces over time. Then the line of "longitude" will represent the total duration in which an intensive change comes to pass, while the line of "latitude" will chart the progressive increase and reduction of the quality. Similarly, if one seeks to determine variations over space, rather than time, one may again trace two lines. In this case, "longitude" will depict the total extension of the surface, while "latitude" will measure the rise and fall of an intensive quality throughout its various sections. Oresme concedes, however, that when the subject of the change is a three-dimensional being, such as a body, its representation poses a special difficulty: no "fourth dimension" can be found in which to illustrate the intensive variation. In such cases, Oresme therefore suggests an expedient. He divides the body into an infinite quantity of surfaces, each of which can be quantified, in the same way, as a line.[30]

Qualities and motions may then be represented in geometrical "configurations." Continuous variations in intensive degree will be illustrated as segments, and the relations between such variations will be, therefore, legible as the ratios of lines. "For whatever ratio is found to exist between intensity and intensity, in relating intensities of the same kind," Oresme explains, "a similar ratio is found to exist between line and line, and vice versa. For just as one line is commensurable to another line and incommensurable to still another, so similar in regard to intensities certain ones are mutually commensurable and others incommensurable in any way because of their [property] of continuity."[31] If one wishes quantitatively to compare the intensive variations of two surfaces or durations, it suffices, then, to measure the areas of their "configurations." Intensive *uniformitas* will be depicted by a prolonged "latitude" in a stable relation to "longitude." In geometrical depiction, two parallel segments will then be drawn, which, when tied at their beginning and end, will compose a rectangle. Intensive *difformitas*, by contrast, will be illustrated by asymmetrical extensions. For qualities in increase and decrease, rising and falling latitudes will be traced, just as, for qualities in continuous progression, there will be curves and slopes. When tied to their respective longitudes, these latitudes will form all manner of geometrical shapes. Indeed, in both diagrams and words, Oresme demonstrates that the ratios of such magnitudes may compose any plane figure, be it rectangle, triangle, or polygon.

Before Oresme, medieval thinkers availed themselves of this system of geometrical illustration in various fields of study, from medicine to theology. Galen, al-Kindi, Avicenna, Averroës, and numerous physicians after them sought to calculate the degrees of elementary qualities necessary for the state of health, and they distinguished them from the disproportions that cause illnesses.[32] Yet

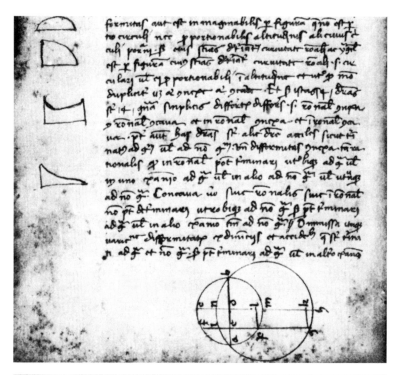

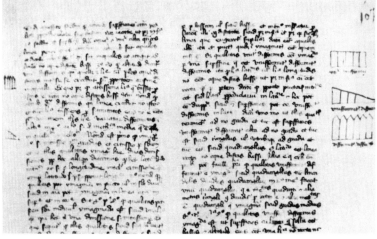

Oresme's Geometrical Configurations (Marshall Claggett, ed., *Nicole Oresme and the Medieval Geometry of Qualities and Motions* [Madison WI, 1968], figs. 9 and 17).

in his *Treatise on the Configuration of Qualities and Motions*, Oresme showed that such studies could be brought to bear on perceptible qualities, such as sound. The second book of his work took as its point of departure one striking claim: that sonorous qualities are in truth of the order of magnitude, not multitude.[33] That principle called into question a thousand-year tradition, which maintained, by contrast, that, to be intelligible, sounds must be essentially arithmetical in kind. To define musical entities as magnitudes, not multitudes, was to overturn the foundations of the art of music. Oresme knew this well. At the beginning of his discussion of acoustical phenomena, he argued that sound possesses two types of intensity: "one in acuity, the other in strength" (*una in acutie, aliam in fortitudine*).[34] In the first of these two distinctions, one may recognize the differential character of sound to which Boethius dedicated most of his doctrine: that of acuity and gravity, later understood in terms of the "height" of pitch. Boethius had assumed that such a character reveals essentially discontinuous units of sound, namely, tones, constitutive of intervals. Representing the acuity and gravity of musical sounds as degrees of intensity, Oresme suggests otherwise, indicating that tones may be essentially continuous.

In naming another variety of acoustical intensity, Oresme departs still further from the arithmetical tradition, as he himself reminds his readers. "Boethius attempts to give the cause of the first kind of augmentation and diminution," Oresme writes, discussing the two qualities of sound, "but he does not speak of the second."[35] The "strength" or "volume" (*fortitudo*) of sounds can be explained only as a quality susceptible to continuous reduction and growth. To those two intensive dimensions of sound, however, Oresme also adds a third, unmentioned—and perhaps unimagined—by Boethius and his commentators. This is the qualitative dimension of sound now commonly called "timbre." Oresme suggests that tones that appear to be discrete may be aggregates, composed of many minor acoustical magnitudes imperceptible to human ears. Depending on the ratios in which these smaller elements are mixed together, different qualities of sound may be produced.[36] Oresme had observed this fact: strings made from the guts of different animals are audibly distinct in their textures, even when the notes sounded are themselves in principle identical. Such phenomena are reducible to intensive fluctuations in the ratios of their constituents. For this reason, no purely arithmetical notation can record them. Like the latitudes of "height" and "strength," timbre may be best transcribed by geometrical configuration.

Oresme did not limit his doctrine of continuous intensities to this world. He pursued the geometrical quantification of qualities to the heavens, considering the idea that qualitative sounds might fill the vault above the earth. In his *Treatise on the Commensurability and Incommensurability of the Celestial Movements*, Oresme proposes to consider astral qualities and motions, like all others, as magnitudes susceptible of configuration and exact measurement.[37] The relation of the velocity of one celestial body to another, he explains, can be considered equivalent to the geometrical ratio of one magnitude to another. Since movements are known to produce sounds, such motions may also be sonorous; the relations between them, moreover, may compose a harmony, albeit one possibly inaudible to us. To be sure, Oresme does not claim to be able to determine astral velocities arithmetically, any more than he wishes to measure magnitude by multitude, reducing the proportions of lines to ratios of numbers. He seeks solely to define the geometrical relations between the starry motions. More precisely, he addresses a single question that arises inevitably once astral velocities are conceived as continuous magnitudes. If the heavenly motions are related as two lines in proportion, their speeds, like segments, may be incommensurable as well as commensurable. In other words, the celestial velocities might share some least common measure. Yet the contrary hypothesis is also admissible: irrationalities may rule the motion of the stars.

Earlier authors had noted that the conjunctions of the stars are by no means absolutely regular; their movements could not be perfectly predicted. In the second century B.C., Theodosius of Bithynia had already raised the problem in a book, *Days and Nights*, in which he sought to understand the reasons why the motions of the sun and moon appear not to coincide.[38] Oresme's older contemporary, the mathematician, astronomer, and musician Johannes de Muris also examined the problem in his *Quadripartitum numerum*.[39] Yet by the second half of the fourteenth century, thinkers could address the question by means of the innovative mathematical rules of Scholastic physics. Aristotle's works on natural philosophy were known to contain a formula for the calculation of speed: velocity, it was said, could be arithmetically derived from the relation of force to resistance.[40] But that principle led to contradictions and even absurdities, as Aristotle himself had acknowledged more than once. As an example, one may consider the case in which force and resistance are equivalent. If one takes the old formula as a rule, one will be obligated to assert that from such a ratio of identity, a certain speed must be derivable; yet of course, in such a case, nothing will move.[41]

In the twelfth century, Averroës therefore revised the formula, relating speed not to force per se, as had been done, but to the excess of force with respect to resistance.[42] Some three decades before Oresme, Thomas Bradwardine had proposed a far subtler solution, known today to historians of science as Bradwardine's "rule" or "function."[43] On the basis of this principle, later thinkers derived velocity from the exponential ratio of force to resistance, where the exponent is itself a ratio.[44] This was to be Oresme's point of departure: a velocity, as he wrote, derives from a "ratio of ratios" (*proportio proportionum*). Considering the matter attentively in several related works, Oresme formulated an original theorem on the mathematical structure of such exponential relations: namely, that from any given sequence of geometrical relations, more irrational than rational "ratios of ratios" can be produced. In other words: exponential "proportions of proportion," more often than not, are arithmetical disproportions—or mathematical irrationalities—that no integers may express. Moreover, Oresme argues, the more numerous the set of possible ratios, the greater the probability that any two of them, placed in exponential relation, will be incommensurable.[45]

From this rule, one might well infer that the heavenly velocities, whatever their exact values, are most probably also incommensurable among themselves. Oresme himself implies as much in his book, *On the Proportion of Proportions.*[46] But in his *Treatise on the Commensurability and Incommensurability of the Celestial Movements*, he declines to offer an unequivocal answer to the question. In the first two parts of the work, he advances numerous scholarly arguments drawn from the fields of mathematics, astronomy, physics, and mechanics. Yet in the third and final part of the book, Oresme consigns the discussion to the world of fable and myth. Philosophical argumentation gives way to allegorical letters, and the work ends in the recounting of a dream. In a vision, Oresme sees Apollo, accompanied by the Muses and the classical fields of knowledge, each figured by an allegorical personage. The ancient god commands the representatives of the mathematical disciplines to enlighten the medieval thinker on the matter that has so aroused his interest. But the reader learns that on this question, the ancient disciplines of quantity are irreducibly at odds with themselves.

In an extended debate, Lady Arithmetic and Lady Geometry take opposing sides, "as if they were litigants," Oresme explains, "in a lawsuit."[47] Lady Arithmetic speaks first. She claims the privilege of being the "firstborn" of all the branches of mathematics, her objects, numbers, being the most noble of all idealities. Boethius, she recalls, taught that "everything that proceeded from the very origin of things was formed with reference to numbers";[48] Macrobius, pointing to her priority with respect to Lady Geometry, declared that "numbers preceded surfaces and lines."[49] Could it be imagined that the heavens not be measurable by such perfect unities? "If I am thus expelled from the celestial regions," she asks, "to what part of the world shall I flee—or am I to be banished beyond the boundaries of the world?"[50] Irregular as well as irrational, incommensurable quantities belong below the moon, as Aristotle wrote, "because things in these lower regions have twisting [and confusing] motions as a result of their remoteness from a proper divinity."[51] Surely, "God would not permit such disorder near himself in the heavens."[52] "Furthermore," Lady Arithmetic remarks, were heavenly motions incommensurable, "our dutiful daughter, sweetly sounding Music, would be deprived of celestial honor, even though she participates in ruling the heavens, as many physical occurrences bear witness."[53] If the motions of the spheres were magnitudes and not multitudes, the skies would fill with ratios "discordant and strange in [their] harmony and, consequently, foreign to every consonance, more appropriate to the wild lamentations of miserable hell than to celestial motions that unite, with marvelous control, our musical melodies soothing a great world."[54]

Yet Lady Geometry is undaunted. She, too, lays claim to mathematical primogeniture. She declares her object, magnitude, to embrace all quantities, number being but one variety of continuity.[55] To the claim that, to be consonant, spheres must move in speeds commensurable with each other, Lady Geometry responds that harmonic sounds result from the relations of not the velocities but the masses of sounding bodies. "This explains," she notes, "why Pythagoras did not measure the motion of the hammering or the force of the blows, but [instead] sought the ratio of the hammers, a quantity which he knew by their weights."[56] The heavenly masses might well be related in such proportions, for "they say," as Lady Geometry concedes, "that the relationship obtaining between the sun and Venus is a *diesis*, a ratio consisting of the numbers 256 and 243."[57] But, she notes, one can hardly know whether such harmonies are audible as well as intelligible. And even if the motions of the stars issued in music, they could never be commensurable, for then they would be unchanging. Eternal asymmetries would be more beautiful:

> What song would please that is frequently or oft played? Would not such uniformity produce disgust? It surely would, for novelty is more delightful. A singer who is unable to vary musical sounds, which are infinitely variable, will no longer be thought best; rather, he will be taken for a cuckoo. Now if all the celestial motions were commensurable, and if the world were eternal, the same, or similar, motions and effects would necessarily be repeated.[58]

Lady Geometry does not hesitate to draw this radical conclusion: "For this reason," she declares, "it seems more delightful and perfect—

and also more appropriate to the deity—that the same event should not be repeated as often, but that [on the contrary] new and dissimilar configurations should emerge from previous ones and always produce different effects."[59]

Awakening keeps the dreamer from witnessing the end of the debate. "Alas, the vision vanishes, the conclusion is left in doubt, and I am ignorant of what Apollo, the judge, has decreed on this matter."[60] But as several scholars have noted, a passing remark contained in the thinker's Middle French treatise *On the Heavens* leaves little doubt as to Oresme's own position on the question: "In a treatise called *On the Commensurability and the Incommensurability of the Celestial Movements*," we read, "I once showed, by various means, that it is more probable than improbable that some of the movements in the heavens are incommensurable" (*ce est plus vraysembable que n'est l'opposite, si comme je monstray jadys par plusseurs persuasions en un traité intitulé De Commensurabilitate vel incommensurabilitate motum celi*).[61] Yet in that treatise itself, Oresme, a man of the Middle Ages, hesitated to write anything so explicit. Although he had demonstrated the likelihood of astronomical irrationalities, he resolved not to proclaim it. Perhaps he recalled the fate of the impious Pythagorean of antiquity, punished by the gods for having disclosed the secret of the incommensurable; perhaps it was in ironic deference to that old revelation that Oresme chose Apollo to act as arbiter in his new vision. Before waking, the author let the god suggest that some reconciliation between the disciplines might be found: "Do not seriously believe," Apollo tells the learned dreamer, "that there is a genuine disagreement between these most illustrious mothers of evident truth."[62] Yet the discord, of course, was not to be surmounted. Irrationalities had been introduced into the perfect movements of the spheres. Oresme had detected a crack in the firmament of the ancient and medieval world; with the tools of mathematical argumentation, he had shown, moreover, that its existence was not only possible but also probable. That fissure could not be undone. In time, it was to grow only more imposing, until the cosmos of the ancients came definitively apart. Lady Geometry was to have her wish: her sister Arithmetic, once the firstborn of the disciplines, would be "banished beyond the boundaries of the world." Her numbers would be dissolved into shapes and solids; her old multitudes would be resolved into the continuities of magnitudes. The moderns would fear no disproportions. In the universe that was to come, "new and dissimilar configurations" were to emerge, in "deformity" as well as in "uniformity," and always to "different effects."

CHAPTER FIVE

Ciphers

The doctrine of modern harmony began as the imitation of the ancient. Gioseffo Zarlino, chapel master of St. Mark's in Venice, has long been taken to be its author. His *Harmonic Institutions* of 1558 ushered in a new age in the quantitative reasoning of sounds. "It is safe to say," it has been noted, "that probably no theorist since Boethius was as influential upon the course of the development of music theory."[1] Yet however innovative it may have been in substance, in form, Zarlino's theory was in the image of Boethius's. In the first part of his monumental treatise, Zarlino declared that "music is a science, which considers Numbers, and Proportions."[2] He upheld the thesis that the domain of numbers and their relations coincides, in truth, with the entirety of the created world. "The things created by God were ordered by Him according to number, and number was the main exemplar in the mind of the maker; therefore, it is necessary that all things that exist separately or together be comprehended by number, and submitted to number."[3] "For this reason," Zarlino remarked, "it is hardly surprising that the Pythagoreans consider there to be something divine in numbers."[4] It almost went without saying that such "numbers" are by definition discrete and divisible, being collections of ones. Yet lest the point be mistaken, Zarlino specified that "every number contains in itself unity several times over."[5] Whether even, odd, prime, quadratic, cubed, or perfect, multitudes were for the author of the *Harmonic Institutions*, as for Boethius, the true objects of *musica*. More exactly, such arithmetical beings pertained to the art of harmony to the degree to which their ratios might be exhibited in the sounds of consonances. After Philolaus, Archytas, Plato, and Boethius, Zarlino naturally recalled the ancient Pythagorean findings: the intervals of the octave, fifth, and fourth may be reduced to the lengths of strings extended in accordance with arithmetical ratios. Nonetheless, for Zarlino, that well-established doctrine was insufficient.

Writing in the mid-sixteenth century, Zarlino could not content himself with the theory of the ancients, even if he still employed its terms and concepts. As early as the fourteenth century, musicians began to reckon with four intervals unknown to the theorists of classical antiquity: the major and minor sixths, as well as the major and the minor thirds. By Zarlino's time, these intervals were increasingly considered not dissonant, but consonant; far from avoiding them, the Renaissance composers made systematic use of both thirds and sixths. Yet the ancient doctrines furnished no tools to define these modern sounds. Pythagoras had shown that with the aid of no more than the first four natural numbers, all the classical consonances could be expressed in arithmetical form: the octave may be reduced to the relation of two to one (2:1), the fifth to the relation of three to two (3:2), and the fourth to the relation of four to three (4:3). The first four integers formed a limited alphabet, which sufficed for the transcription of the consonances known to Greek and Roman musical theory. Yet no relation of these integers could represent the lengths of strings necessary to produce the major and minor thirds and sixths. Had Zarlino been of a different temperament, he might well have chosen, for this reason, to reject the old rules of *musica* outright, since in their classical form, they were plainly inadequate to the harmony of the moderns. But in his *Institutions*, Zarlino famously opted otherwise. He demonstrated that one may write the sounds of the early modern age in classical form if one effects one modification: the letters of arithmetical transcription must be expanded, and one must learn to count from four to six. Then, the theorist showed, the major third may be reduced to the relation of five to four (5:4) and the minor third to that of six to five (6:5). Similarly, the major sixth may be reduced to the relation

of five to three (5:3), and the minor sixth to that of two times four
to five ([2]4:5).

On the surface, those transcriptions recalled the findings of
Pythagoras. The *Harmonic Institutions* marshals much evidence
in support of this impression. Drawing on many ancient sources,
Zarlino argued at length for the classical foundations of his doctrine
of the musical *senarius* or *senario*, as he baptized it.[6] He noted that
the number six has always been known to be exceptional in form.
Astronomy is familiar with six planets (the moon, Mercury, Venus,
Mars, Jupiter, and Saturn), and six circles mark the heavens (Arc-
tic, Antarctic, the Tropic of Cancer, the Tropic of Capricorn, the
Equinoctial, and the Equinox). Ancient philosophy reckons with six
"natural offices" (size, color, shape, interval, state, and motion), and
it admits six ages of human life. Mathematicians teach that six lines
define the triangular pyramid, and they know that six sides compose
the surface of the cube.[7] The evidence, he showed, is copious. But
Zarlino also argued for the special distinction of the multitude of
six in itself. Six, as he reminded the reader, is the first in the series
of perfect multitudes. It is, in other words, the first number to
constitute the sum of all the factors into which it may be resolved
($1 \times 2 \times 3 = 1 + 2 + 3$). Hence, Zarlino claimed, its essential harmony.

Such varieties of proof are less than absolutely compelling.
Doubtless, other numbers are equally symbolic. As H. F. Cohen
has noted, "it is not at all difficult to carry out a similar exercise
with, for instance, the number *seven*."[8] Moreover, Zarlino's numbers
were—at least in one decisive case—less than adequate to the sounds
whose values they aimed to define. That one case is the minor sixth.
The *Harmonic Institutions* presents this consonance as defined by the
relation of "two times four, to five." But the truth is that such a ratio
is shorthand for another: eight to five (8:5).[9] Tethered to his *senario*,
Zarlino could not transcribe this interval in its simplicity. Yet he also
could not have ventured in his count as far as the number eight, since
to do so, he would have had to pass through the number seven, and
then he would have had to confront a set of ratios that, while arithmetically coherent, correspond to no musical consonances. Intervals

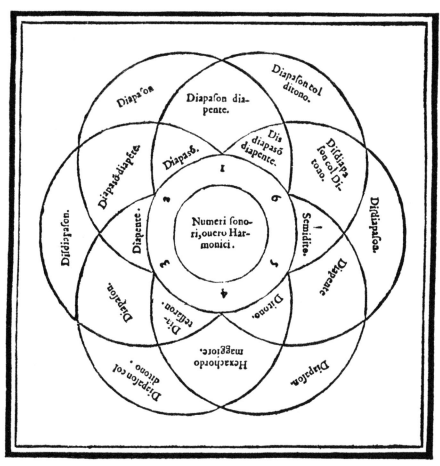

Zarlino's *Senarius* (Gioseffo Zarlino, *Le Istitutioni harmoniche* [New York, 1965], p. 25).

such as the augmented fourth and the diminished seventh—both unacceptable in Renaissance harmony—would have been difficult to set aside.[10] In his wish to transcribe the musical world of the moderns in an idiom resembling that of the ancients, Zarlino thus chose to rise to six, but to no more. He resolved the question of the minor sixth by arguing that it is contained in the *senario* not actually, but potentially, as eight is a potential form of four and two. One suspects that the metaphysical solution, while ingenious, belied a difficulty, the price the *Harmonic Institutions* paid for its compromise of old and new. Modern consonances might be written by numbers similar in form to the ancient, yet only on the condition that such multitudes changed in quantity and in substance, becoming, as it were, the ciphers of their old selves.

Soon, however, the foundation of all such considerations gave way. From Boethius to Zarlino, a long tradition of thinkers held musical sounds to be arithmetical in nature. Pythagorean theorists of harmony held that for every consonance, a single arithmetical ratio could be found; they taught, indeed, that the beauty of intervals was that of "harmonic numbers" sounding together. Despite its imperfection, the physical world could thus be seen to partake of the ideal, since in music, physical occurrences could be led back to ratios of essences. Franchinus Gaffurius's *Theorica musicae*, published in 1492, still retold the tale of the discovery of this doctrine: Pythagoras once stumbled into a forge in which he discovered five smiths working with five hammers, four of whose weights, being in simple proportions, necessarily produced a set of pleasing sounds.[11] Whatever the worker, whatever the instrument, and whatever its matter, certain arithmetical regularities were bound by nature to produce certain consonances. Yet in the mid-sixteenth century, an Italian author announced that it was not so. He was Vincenzo Galilei, an erstwhile pupil of Zarlino's and a specialist of lutes. In a *Discourse Concerning the Works of Sir Gioseffo Zarlino of Chioggia*, published in Florence in 1589, Galilei boldly announced that in connection with the doctrine of Pythagoras, he would now reveal to the world "two false opinions of which men have been persuaded by various

Jubal, Pythagoras and Philolaos (Franchinus Gaffurius, *Theoria musice*, 1492)

writings, and which I myself shared until I ascertained the truth by means of experiment, the teacher of all things."[12]

The "two false opinions" both involved the relation of number to sound. Boethius and his followers had taught that Pythagoras discovered the regularities of consonance and ratios through the perception of sounding mass. The Sage was said to have heard harmonies in the percussions of hammers. Later, having returned home, he would have continued his work with strings. And just as the weights of the four hammers heard in the forge were in the relation of six, eight, nine, and twelve, so the respective cords with which Pythagoras experimented would be in length six, eight, nine, and twelve. Woodcuts printed in Gaffurius's *Theorica musicae* illustrated these equivalences. Several portraits showed Pythagoras reckoning with sets of hammers, bottles, and weights, upon all of which were clearly inscribed the same sets of integers. Such images testified to the first "false opinion" that Vincenzo Galilei corrected. Two strings whose lengths stand in the relation of two to one will produce an octave, but, Galilei discovered, objects whose weights are in that ratio will not. For masses to produce an octave, they must be in the relation of not two to one (2:1), but four to one (4:1); for weights to produce a fifth, they must be in the relation of not three to two (3:2), but nine to four (9:4); and for weights to sound the fourth, they must be in the relation of not four to three (4:3), but sixteen to nine (16:9). In short: the harmonic proportions derived from strings must be squared for them to hold for mass.[13]

The second false opinion that Galilei set out to dispel followed from the first.[14] The Pythagoreans held that only those arithmetical relations that produce consonances when applied to string lengths might be considered to be harmonic. For the ancient musical theorists, such relations were exclusively of two kinds. They could, first, be "superparticular": in such inequalities, a greater term exceeds a lesser by one part of the lesser. The fifth (3:2) and the fourth (4:3) are of this variety. And, second, in the classical doctrine, harmonic relations could be "multiple" in form: in this case, the lesser term constitutes a factor of the greater term. Examples of such inequalities include the single, double, or triple octave (2:1; 4:1; or 8:1). While admitting the existence of new consonances, Zarlino still sought to maintain this ancient theory, despite the considerable difficulties it caused him.[15] Yet by reference to the quadratic proportions of weights, Galilei demonstrated that inequalities of radically nonclassical form could also produce consonances. It sufficed to consider the relative masses necessary for the production of a fifth (9:4) or a fourth (16:9). Neither inequality is "superparticular" or "multiple," yet each, when sounded in weights, gives rise to indisputably harmonic intervals.

Today, Galilei's discoveries may seem of relatively small consequence. To be sure, they did not refute the ancient idea of a link between number and sound. They merely demonstrated that the inequalities required for the production of intervals in strings are not identical to those required for intervals in mass. But the implications of this apparently simple finding were, in truth, immense. Galilei had loosened the bond that tied musical sounds to immutable arithmetical ratios. He had, in other words, disproved one basic axiom of Pythagorean harmonics: that the intervals of music are the expressions of unchanging ratios of essences. The ancients and medievals who followed the teachings of the master of Samos believed themselves to perceive in music "sonorous numbers" (numeri sonori).[16] Galilei showed there could be no such things, for the inequalities that produce a consonance in one material may be at odds with those that produce it in another. Consonance, then, could no longer be viewed as the sound of stable multitudes. The mathematical ratios of harmony now appeared as no more than the signs of different types of beings, signs whose meaning, therefore, depended on material. "Two to one," "three to two," and "four to three" no longer named ideal relations, perceptible in this world despite its constant flux of generation and corruption. Such expressions would henceforth be the signs of measurements, whose sense depended on the various types of bodies to which they were referred.

It would be difficult to overestimate the consequences of this discovery. Vincenzo Galilei, admittedly, did not unfold them all himself, and at least a portion of his new propositions on the harmonic properties of objects was unsound.[17] Yet in retrospect, it is clear that his explosion of the idea of "sonorous number" announced a fundamental change in the definition of harmony as a field of study. Pythagorean thinkers of the ancient and medieval periods had defined consonances by numbers because they believed such sounds to be, in essence, mathematical. But they never claimed as much of all physical beings. The ancient and medieval cosmologies, at least in their Aristotelian forms, were hierarchical in structure, and they distinguished between beings of many types. At the summit of the scale of Being lay the eternal spheres, whose perfectly circular movements could be precisely defined because they were mathematical. At the lowest level, there were corruptible bodies, in themselves inconstant and therefore uncertain. Between these two orders of Being, music effected a conjunction: thanks to the doctrine of harmony, eternal ratios could be perceived in sounds. In the age that began with Galilei, however, the grounds for that conjunction would give way. When the stratified cosmos of classical *epistēmē* came to be supplanted by the homogenous universe of modern science, no being, in itself, would be identifiable with a number or with a relation of numbers.

That principle finds a precise correlate in the early modern philosophy of mathematics. It may be simply stated: no number or relation of numbers, in itself, would be identifiable with a being. Such a thesis was at odds with the traditional understanding of arithmetic as the study of the relations of essences. For the classical thinkers, the theory of *arithmoi* was, in itself, a doctrine of arithmetical beings, "numbers" being definite collections of definite things. As Alexander of Aphrodisias argued, "every *arithmos* is the *arithmos* of something" (πᾶς γὰρ ἀριθμός τινός ἐστι), where the "something" in question, moreover, is definite and discrete.[18] With the rise of a new theory of mathematics in the sixteenth and seventeenth centuries, however, these ideas became obsolete. As Jacob Klein has shown in detail, François Vieta, Simon Stevin, John Wallis, and Descartes conceived the philosophical foundations for modern algebra in formulating a fundamentally new metaphysics of mathematical objects. The modern thinkers defined "number" as the symbol of an idea of any given size. Thus Vieta, Klein writes,

> preserves the character of the "*arithmos*" as a "*number of...*" in a peculiarly transformed manner. While every *arithmos* intends immediately the things or the units themselves whose number it happens to be, his letter sign intends directly the general character of being a number which belongs to every possible number, that is to say, it intends "number in general" immediately, but the things or units which are at hand in each number only mediately.[19]

The new conception can be expressed in the terms of the old Scholastic distinction between two types of "intention": the *intentio prima*, which signifies a thing, and the *intentio secunda*, which signifies an intention itself, such as an idea. For Vieta, a number primarily signifies the object of a second intention: the concept of a "general number," a "species," or purely conceptual formation.[20] Similarly, Stevin, Wallis, and Descartes all understood numbers to be not ideal essences but symbolic notions to be employed, as in algebra, for calculation.[21]

In a telling passage, Stevin explains that the ancient Greeks failed to arrive at a correct definition of number because they lacked a symbolic notation. They suffered from "the absence of necessary equipment, namely ciphers" (*faute d'appareil nécessaire, nommément des chiffres*).[22] The allusion to the "cipher" is to be taken seriously. The sixteenth-century authors were well aware that the ancients had held the smallest number to be two, on the grounds that one was the first principle of unity, or monad, of which all numbers were collections. In such a world, zero could clearly have no place in the doctrine of arithmetic. This changed in modernity, when the Europeans received from medieval Arabic mathematics the idea of the zero (*sifr*), the "cipher" for nothing, represented by a typographical point or small circle.[23] In the modern age, the zero was not only to be a "number"; for many thinkers, the zero, rather than the "one," became the very principle of numbers. Stevin thus writes that the arithmetical "point of number" (*poinct de nombre*) corresponds precisely to the geometrical "point of size," for just as the line may be reduced infinitely toward a point, so any number can be decreased to zero.[24] Such a conception could not but bring about a transformation in the definition of arithmetic. Ancient mathematics had opposed geometry to arithmetic as a knowledge of the continuous to that of the discontinuous. That distinction now became defunct. "Number is not at all a discontinuous quantity," Stevin argues:[25]

> As a continuous body of *water* corresponds to a continuous *wetness*, so a continuous *magnitude* corresponds to a continuous *number*. Likewise, as the continuous wetness of the entire body of water is subject to the same division and separation as the water, so the continuous number is subject to the same division and separation as its magnitude, in such a way that these two quantities cannot be distinguished by continuity or discontinuity.[26]

In the seventeenth century, Wallis took one step further. Since the moderns admit the continuity of units, he argued, they may also consider numbers to be fractional (*fracti*), irrational (*surdi*), or algebraic (*algebraici*). Naturally, were the numbers in question classical multitudes, such determinations would be unthinkable; *arithmoi* must be discrete and definite. Yet numbers, Wallis explains, are not the names of ones; they are signs of ratios. "And so also the whole of arithmetic itself," he writes,

> seems, on closer inspection, to be nothing other than a theory of ratios, and the numbers themselves nothing but the 'indices' of all the possible ratios whose common consequent is 1, the unit. For when 1 or the unit is taken as the reference quantum, all the rest of the numbers (be they whole, or broken, or even irrational) are the indices or exponents of all the different ratios possible in relation to the reference quantum.[27]

As Klein observes, "here 'number' no longer means 'a number of...'; rather a number now *indicates* a certain ratio, a *logos* in the sense of Euclid."[28] The distinction between multitude and magnitude, arithmetic and geometry, comes undone. Numbers emerge as the symbols of any possible proportion. No more striking example can be found than the cipher of ciphers, zero itself. For Wallis, it constitutes the sign of a ratio, like all others: in a system of formal notation in which a given quantum is 1, for example, 0 will simply constitute the sign for the privation of that quantum. Yet if such a sign represents a number, then, by definition, numbers cannot always signify beings, since nothing that is and that remains itself may be said to be equivalent to zero. The link between the elements of arithmetic and metaphysics no longer holds.

The age that opened with an imitation of the Pythagorean decipherment of sound thus led, within a century, to the integral transformation of its letters. From four harmonic numbers, Gioseffe Zarlino passed to six, before Vincenzo Galilei showed that no ratio of numbers could imply a single sound, as no number, for the modern philosophers, could constitute an ideal essence in itself. One hundred years after Zarlino's *Harmonic Institutions*, the axioms shared by music and arithmetic, in place for two millennia, grew obsolete. The basic elements of harmony and mathematics would henceforth be disjoined: entities would not be reducible to integers; integers would not be entities. Between the physical reality of sounds and the symbolic notions of mathematics, an abyss had opened, which no appeals to classical authority could conceal.

Yet the partition of metaphysics and logic announced a new possibility. Precisely because the modern physical universe, unlike the ancient cosmos, was not in itself mathematical, it could be measured to an unprecedented extent. In the ancient and medieval world, some beings alone had been measured in intelligible quantities; all others were held to be too uncertain to be apprehended by exact calculation. While the cycles of the spheres, for example, were thought to be perfectly circular, motions in the sublunary realm were believed too variable to be defined with mathematical precision; while consonances were taken to be arithmetical, dissonances were believed to be, by nature, incapable of being grasped as such. In the new physical sciences of nature, however, such distinctions no longer held. Any object, if empirical, might become an object of experimental investigation. States of rest and motion, qualities and quantities, Becoming no less than Being—all could be submitted to the new methods of the emerging natural philosophy. What was not mathematical in itself, in short, could be "mathematicized."[29] The domain of sound was no exception. From Galilei's son to Isaac Beeckman, Marin Mersenne, Christiaan Huygens, and Gottfried Wilhelm Leibniz, the next generations would show that acoustical phenomena could be precisely measured in the quantities of modern science. After the age of ancient *epistēmē*, modern science would effect a new conjunction. Sounds shorn of ideality would be tied to ciphers without substance, and, through that articulation, the order of a new nature would be transcribed.

CHAPTER SIX

Temperaments

In 1585, the distinguished Venetian mathematician and scientist Giovanni Battista Benedetti published one *Book of Diverse Mathematical and Physical Speculations* (*Diversarum speculationum mathematicarum et physicarum liber*).[1] Among the materials contained in this work are two undated letters addressed to the composer Cipriano de Rore, who appears to have turned to Benedetti for clarification concerning the nature and properties of sounding instruments. Benedetti, himself also a musician and composer, was particularly qualified to respond to the various questions posed to him. Today, it is clear that his letters constituted far more than a divulgation of findings familiar to scientists of his time. In the second of the two epistles, in particular, Benedetti proposed an account of harmonic consonances that, in hindsight, can be said to have announced the nascent field of the mathematical study of sound, or "acoustics." Benedetti tacitly set aside the ancient and medieval practice of defining musical consonances by the measured lengths of sounding strings. In its place, he proposed a method that allowed sounds to be quantified by the numbers of the modern epoch. Perhaps for the first time in the history of the study of the laws of sounding bodies, Benedetti correlated sounds not with the lengths of strings but with their vibrations. A new age in the transcription of nature by number now began.

In his second letter, Benedetti stated three closely related physical principles. The first had its root in a fact of which "everyone," as Benedetti noted, was well aware, namely, that "the longer the string, the more slowly it is moved."[2] Hence the rule that Benedetti now expressly affirmed: the magnitudes of string lengths and vibrations are inversely proportional, or, as Claude V. Palisca has observed in more modern terms, "the ratio of the frequencies of two strings varies inversely with their lengths, the tension being equal."[3] Second, Benedetti established that the consonances of harmonic intervals derive from coincidences in the termination of vibratory cycles. In other words, if two strings cease to move at the same moment, then their sounds will be concordant; inversely, if the two sounds of two bodies are harmonious, it is because their vibration has ceased simultaneously. If two strings sound the interval of the octave, for instance, it is because the movements of the two strings come to a standstill in moments that are coincident. Each time the longer string ceases to vibrate, the shorter one will be motionless; inversely, every other time the shorter string ceases to move, the longer one, too, will be immobile. Benedetti had intuited that all harmonies derive from one periodic event in the time of sounding bodies: the simultaneous endings of their movement.

Finally, Benedetti suggested that such findings furnish elements with which one may define degrees of consonance, stretching from unison to all the other intervals. An index of harmonic sounds may then be obtained. To reach its terms, two simple steps are necessary. One must consider the number of vibratory motions made by two strings in a given moment; then, one must multiply the two numbers. Two strings will sound in unison if they vibrate the same number of times, ceasing their motions simultaneously. One may therefore assign to the sound of unison the value of one times one (1×1), or one (1). Two strings will sound the octave if, for every vibration of the first string, the second string vibrates twice. One may therefore attribute to this consonance the value of two, produced by one times two $(1 \times 2 = 2)$. In his second letter to Cipriano de Rore, Benedetti shows that each of the traditional consonances of musical theory may be subjected to such a mode of representation. The old ratios of string lengths may then be rewritten as ratios of string vibrations. Just as the sound of the octave can be derived from the relation of two vibratory cycles to one, so the fifth may be tied to the relation of three vibratory cycles to two. And just as one may assign to unison the number "one," so one may attribute to the interval of the fifth the number that results from the product of two and three, that is, six; similarly, one may represent the fourth as four times three, or twelve. Gradually, an arithmetical scale of musical concordance in vibration will emerge. Simple consonances may be correlated with single numbers, and numbers, Benedetti notes, will in turn "correspond to each other in a wondrous proportion."[4]

Benedetti lent particular weight to the index of concordance he established.[5] He dwelt little on the first and second principles he so clearly affirmed, and although he may well be taken to be the first thinker to have defined consonance by vibration, he offered no demonstration of the relation of harmonic sounds to cycles of motion. A generation later, Galileo Galilei, who knew Benedetti's works on mechanics and acoustics, also maintained that musical ratios can be reduced to ratios of vibration. In a passage of the *Discorsi*, which he published in 1638, Galileo not only affirmed that consonances could be derived from the coincidence of pulsations but also sought to prove this fact.[6] Most historians of science, however, attribute the first rigorous demonstration of the correlation of quantities of sound and quantities of vibration to the Dutch scientist Isaac Beeckman (1588–1637). A journal entry dated to 1615 contains a simple, yet decisive geometrical proof that the frequency with which a string vibrates stands in direct proportion to the acuity of its sound.[7] Being himself an atomist, Beeckman held that tones, like all other things, consist of small and indivisible beings "at diverse distances from each other, that is, at intermediate empty spaces,"[8] set in movement by vibration and carried, with varying force, from the place of their first motion to the faculties of human hearing.[9] His contemporaries might well refuse such an atomistic physics; they might hold theories, for example, in which sounds were said to result from the motions not of discrete "corpuscles" but of waves. Yet the principle Beeckman demonstrated would remain, and gradually it would be accepted that sounds could be defined by those ratios of vibrations (or beats or pulsations) to time still today called "frequencies."

It is worth pausing to consider the relation these acoustical ratios bear to the older musical inequalities of the tradition of Pythagorean harmonics. As Benedetti suggested in his letter, the idea that the acuity of a string's sound is in inverse relation to its length was not entirely novel. Ancient and medieval authors had also not failed to hypothesize that sounds might constitute the audible results of vibrations. It is not wholly inconceivable that a classical student of sensible things might have recognized that when two strings produce tones differing by an octave, one string must vibrate twice as fast as the other. Yet one might also imagine that for the classical and medieval thinkers, the fact of consonance between strings would have been more likely explained by reference to the regularities of the monochord and, more precisely, to its measured lengths. Were it to be discussed at all, the sound of vibration would have been considered with a view to the principle that musical intervals may be correlated with the numbers measuring the extensions of strings. In 1589, Vincenzo Galilei had shown that such correspondences depend on the nature of the material that is measured, the relations that can be relied upon for string length being inapplicable, for example, to the consideration of bodies such as bells. The finding that Benedetti announced in those same years reestablished a relation between number and sound of the kind Vincenzo Galilei had called into question. With the discovery of the correspondences between vibrations and sound, an unequivocal link, once more, could be established between numbers and nature. Ratios, for a second time, could notate the many properties of sounding bodies.

Between the new harmonic reasoning of the world and the old, between the inscription of *arithmoi* in the cosmos and the mathematicization of the universe, there are, however, fundamental differences. These involve both what is numbered and what accomplishes the numbering—both the world, so to speak, and its measures. Several points are worthy of note. The ancient and medieval harmonic *arithmoi* designate quantities of strings; by contrast, the modern numbers quantify vibrations of bodies in relation to time. *Arithmoi*, moreover, serve to define consonances in distinction to all other sounds. For the ancient and medieval thinkers, musical ratios possess positive traits that distinguish them from all other inequalities, thus limiting the field of harmony.[10] In principle, Benedetti's scale of concordance, however, respected no such partition. While he focused in his letter on the consonances accepted by the musicians of his time, his method of quantifying intervals in a single harmonic index could easily be extended beyond the closed set of unison, octave, fifth, fourth, thirds, and sixths. As Palisca has remarked, if one studies the agreement of coterminations of vibrations, "there is no sudden falling off of this agreement of waves when the bounds of the mystic number six are overstepped. Nor is any clear boundary discernible anywhere in the infinite series of musical intervals."[11] Benedetti's analysis offered no clear means to distinguish musical from nonmusical sounds. Composers might invoke the old distinction, but modern science could not easily justify it. Finally, the *arithmoi* of antiquity are by definition "multitudes," collections of one; their perceptibility in sounds points to the presence, in nature, of intelligible forms. The numbers that the moderns employ in their knowledge are of another kind. Ciphers, they are the symbols of any quantities whatsoever, be they discrete or continuous.

Within a hundred years, that fact was to prove decisive. Once sounds could be defined by frequency and once frequency, in turn, could be measured in the new numbers of modern science, one classical limitation in the study of sound gave way. It was no longer axiomatic that musical sounds be identified by rational quantities. As the branch of mathematical physics concerned with the nature of sonorous bodies, acoustics could consider incommensurable as well as commensurable ratios. In the field of music, such irrational quantities were many, as had long been known. The ancient Greek and Latin scales had been adjusted according to the so-called "Pythagorean" tuning system, in which the determining interval is the perfect fifth.[12] Such a system runs up against one fundamental physical limit: in a perfect cycle of fifths, the interval of the octave, by necessity, will be acoustically impure, because the relations of the fifth and the octave are incommensurable among themselves.[13] This fact is often expressed in modern musical parlance by the formula according to which "the cycle of fifths cannot be closed," which means simply that if one extends a fifth from one discrete pitch to another and continues doing so indefinitely, one will never pass through any two pitches that sound an octave or number of octaves among themselves.[14] Perfect fifths simply cannot be reconciled with the consonance of the octave. Yet other perfect musical relations are no less incommensurable among themselves. In many cases, the concords of so-called "triadic music"—octaves, fifths, and thirds—are incompatible in their pure forms. Three major thirds, for example, will by nature fall short of composing an octave, even as four minor thirds will always exceed it. Such divergences are inevitable. A physical and mathematical law dictates this rule: not all musical sounds may be strictly reconciled with each other.[15]

The term "strictly," however, points to a further fact of capital importance: though many intervals are incommensurable in their pure forms, they may also be played impurely. Then, the discrepancies between them grow less audible to the ear, and the incompatibilities between intervals can no longer clearly be heard. "Temperament" is the name traditionally given to the activity whereby a musician lessens the exactness of musical consonance, thereby diminishing, in a single gesture, both the purity of concord and the perceptibility of a real discord. For example, given the incommensurability of perfect fifths and acoustically pure thirds, one may choose to "lower" one's fifths slightly, performing them impurely, yet rendering one's thirds noticeably more pleasing to the human ear.

Such techniques are doubtless ancient. They may date back as far as to the time of Aristoxenus, who rejected Pythagorean ratios in favor of equal tones and supposed semitones, perhaps to reduce the clash of pure octaves and perfect fifths.[16] In the later Middle Ages, however, the problem of incommensurable relations acquired a new urgency, since musical composition now admitted not only octaves, fifths, and fourths, but also both major and minor thirds and sixths. Were performers to produce these four new intervals without sounding their discords, they had little choice: they must learn to "temper" their harmonic relations. Yet in the Middle Ages, the practice of temperament per se did not become the object of any theoretical investigation. There is a good reason for this fact, which derives from the classical definition of music as the study of multitudes in relation to each other. Grasping essences that are discrete and multiple, ancient and medieval "numbers" apply exclusively to rational inequalities. Such arithmetical entities, for this reason, could not be employed in any art of temperament. Only symbolic formations that represented magnitudes as well as multitudes, irrational and rational relations, might measure the irreducible conflicts of intervals. Only new numbers, the ciphers of modern science, could be employed to calculate how best to diminish and dissimulate the acoustical discords of musical relations.

From the late fifteenth to the sixteenth, seventeenth, and eighteenth centuries, various systems of temperament gradually emerged. Each responded in its own way to the difficulties that musicians faced in reconciling harmonic intervals with each other. Yet four general types of harmonic adjustment may be distinguished, as J. M. Barbour argued in his groundbreaking study, *Tuning and Temperament*. The first solution can be simply summarized. It consists in retaining the Pythagorean system of tuning while systematically modifying its basic interval. Instruments may then still be tuned through the establishment of a cycle of fifths, in the classical and medieval manner, but each fifth is altered, always in the same form, and audibly incommensurable intervals, as a result, grow apparently commensurable. This is the system now known as the "mean-tone" or "mesotonic temperament." As a musical practice, it appears to date back to the last years of the fifteenth century. Thus, in his *Practica musicae* of 1496, Gaffurius notes that organists modify their fifths by "a small, indefinite amount of diminution named temperament (*participata*)."[17] But the details of such a system were not systematically proposed until 1523, when the Florentine theorist Pietro Aron first explained such a system in a chapter of his *Toscanello in musica*, "Temperament (Participation) and the Way of Tuning the Instrument."[18]

A second remedy for the incommensurability of musical intervals consists of altering some relations alone, while allowing others to retain their acoustical purity. In the harmonic spectrum of a single instrument, tones will then stand in various degrees of harmonic purity with each other. Numerous seventeenth-century and eighteenth-century musicians and theorists proposed such "irregular temperaments." The most famous among them was perhaps Andreas Werckmeister, whose 1691 *Musicalische Temperatur* described six distinct systems of harmonic adjustment.[19] Performers playing instruments tempered in such ways will have at their disposal intervals of varying kinds. Some will be relatively pure; others will sound aberrant. Beyond the range of several reasonably pure fifths, for instance, the player will encounter that exceptional, yet unavoidable tone that the Baroque tuners called the "wolf": the one note that must pay the price for the relative purity of all the others.

A third technique consists in the division of the entire octave into twelve half tones of equal acoustical magnitude. That is the method of equal temperament. Already imagined in the first half of the fifteenth century, this system came to be defined with precision in the seventeenth century, when scientists found the exact irrational magnitude that must be employed for each half tone in such a temperament: the irrational magnitude, that is, of twelve times the root of two.[20] Such a calculation could hardly have been accomplished by premodern arithmetic; it was, by definition, the fruit of the symbolic use of numbers as ciphers.

The mathematical arts of the early modern age also gave rise to a fourth variety of temperament, which rested on arithmetical calculations far subtler still. This is temperament by "multiple division." Such a system divides the single octave into more than twelve equal tonal units. As early as 1577, Francisco Salinas argued for such a temperament, proposing that the octave be divided into nineteen tones.[21] In his treatise *The Harmonic Cycle* of 1691, Christian Huygens demonstrated, with greater exactitude, that the acoustical domain of the octave could be decomposed into thirty-one equal elements.[22] At the beginning of the eighteenth century, Joseph

Sauveur, mathematician and physicist to Louis XIV, went further. With an ingenious use of logarithms and an insight into the existence of overtones, he showed how one might well divide the octave into forty-three or three hundred and one parts, of which each interval would constitute a certain number.[23]

All four types of temperament sought to define musical phenomena. Furnishing new mathematical models to supplement ancient practices of tuning, they allowed the incommensurability of intervals to be diminished and concealed. Yet the significance of such systems far outstripped the domain of the art of sounds. The emergence of the theory of temperament marked one greater development. What had once lain beyond the grasp of the harmonic decipherment of the world could now be represented and manipulated by novel mathematical means. In principle, the new arithmetical reasoning of the world was therefore unlimited, and this was perhaps the greatest of the innovations of the tempered universe. The classical and medieval thinkers held the world to contain domains that, in essence, could not be known by any scientific means: regions of intrinsic obscurity. Now, it seemed, such cloudy areas might soon be dispelled, just as any interval, commensurable or incommensurable, could be precisely set in relation to all others. Admittedly, it might be that the book of nature would remain partially illegible to us. Yet the emergence of the theory of temperament suggested that even beyond the threshold of our limited perception, acoustical phenomena might still be calculated. In at least one domain—that of sound, namely, represented by tonal differences—the universe might well be arithmetically intelligible: such was the promise of modern harmony.

Mathematician and metaphysician, early inventor of the calculus and occasional theorist of temperament, Leibniz devised the doctrine of this harmony. "Music," Leibniz wrote in a letter to Christian Goldbach dated April 17, 1712,

> is an occult practice of arithmetic, in which the mind is unaware that it is counting. For in confused or insensible perceptions, the mind does many things that it cannot remark upon with a clear apperception. In

effect, one would be mistaken in thinking that nothing happens in the soul without the soul itself being aware of it. Therefore, even if the soul does not have the sensation that it is counting, it nonetheless feels the effect of this insensible calculation, that is, the pleasure that results from the consonances, the displeasure from the dissonances. From many insensible coincidences, pleasure, in fact, is born.

Yet Leibniz also assigned a role to insensible "non-coincidences," for he granted incommensurable relations parts to play in the simultaneously physical, arithmetical, and psychological order he evoked. "I do not think that irrational relations in themselves please the soul," Leibniz continued, "except when they are slightly distant from rational relations, which are pleasing. Sometimes, however, these dissonances are accidentally pleasing; they come between delights like shadows in order and light, so that we appreciate order all the more."[24]

A century earlier, the notion that musical intervals might be reducible to a rate of vibrations per instant had struck Descartes as ingenious, yet improbable. When Marin Mersenne sent his friend the table of musical consonances that he had developed on the basis of Beeckman's theory of the regularity of corpuscular beats, the father of modern philosophy expressed grave uncertainty about the value of any such considerations for the study of the human faculties of acoustical perception. "Your way of examining the quality of consonances," Descartes wrote, "is too subtle, at least if I dare to judge it, to be distinguished by the ear, without which it is impossible to judge the quality of any consonance, and when we judge them by reason, this reason must always presuppose the capacity of the ear."[25] There was a simple reason why the Cartesian theory could not admit that the perception of musical consonances might derive in some way from the sensation of a multitude of minor beats or vibrations: such movements could not be said to be the objects of any conscious representations, and these were the only ones that Descartes would grant. In his definition of music as an "occult practice of arithmetic, in which the mind is unaware that it is counting," Leibniz, however, reconciled science and psychology, proposing a doctrine of perception in accord with Benedetti's and Beeckman's rule.

Leibniz drew from modern acoustics the elements of a new theory of the mind. With a certain grasp of the fundamental incommensurability of musical intervals, he acknowledged that an absolutely pure tuning of all musical intervals is impossible. As he once explained in a letter to Conrad Henfling, "since incommensurabilities do not allow one to retain complete exactitude, one needs convenient equivalences, and there is a genius in finding them. Our mind seeks even the simplest commensurable, and it finds it in Music, though those people who do not know this do not consciously perceive it."[26] For Leibniz, temperament, therefore, was a necessity. "Like you," Leibniz explained to his correspondent, the inventor of a system of harmonic adjustment little known today, "I believe that this science has not been sufficiently established and cultivated."[27] With some skill and luck in "finding," one might succeed in determining the "simplest commensurable" harmonic relations. Such relations, by definition, would not be commensurable in themselves. Yet they would be the most commensurable, given acoustical incompatibilities. They would be, in other words, the best of all possible harmonic relations, granted that not all such relations can be physically and mathematically commensurable.

Such "convenient equivalences" would be reached by calculation alone, and in his letters to Henfling, Leibniz considered how this might best be accomplished. He granted from the start that no one could perceive such equivalences consciously. They would be apprehended by the mind solely as "small perceptions" (*petites perceptions*): minor affections that the mind represents without any distinct awareness of doing so. Among the many minute and rapid operations carried out by the mind in the absence of any "clear apperception," there would, then, be an unthinking calculation. We would perform an "occult arithmetic," determining the numbers of vibrations and judging, without any awareness, the relations between coincidences of beats. From the unconscious sensation of the commensurability of vibrations, the conscious awareness of pleasure in listening would be born. But Leibniz's remarks suggested a further and less obvious fact: namely, that from the sensation of incommensurable relations, too, an enjoyment may be derived. Although irrational relations "in themselves do not please the soul," in certain arrangements, they may still produce quite considerable and conscious delights. When "dissonances" come between consonances "like shadows in order and light," they are "pleasing" by accident; when they are "slightly distant from rational relations," they are, moreover, regularly "pleasing" in themselves. Incommensurability, while unresolved, then seems to cause an intimation of its own. It is as if, in such moments of sweet dissonance, we felt that the irrationality of certain acoustical relations cannot fully be avoided, sensing that in the best of temperaments, irrationality, by a secondary necessity, must also be for the best. "We appreciate order all the more."

In such a theory of the harmony of sounds, our pleasure therefore plays a crucial role. It testifies to a correlation between our perception and the nature of the world. When, without awareness, we calculate the complex relations between vibrations, dimly sensing the "simplest commensurable" that we happily find in "Music," we perceive an order that may also be fully represented by the numbers of arithmetic. We do not know, then, what we hear, but we may infer that it can be known as such. For Leibniz, the delight in hearing is in this sense exemplary of one great principle: there is no pleasure of the senses that is not reducible to a pleasure of reason and that cannot be referred, in turn, to the perfectly intelligible design of the universe. "Music charms us," Leibniz explains in the "Principles of Nature and Grace, Founded on Reason,"

> even though its beauty consists only in the concord of numbers, and in the counting of beats or vibrations which coincide in certain intervals, a counting of which we are not aware and which the soul never ceases to perform. The pleasures that sight finds in proportions are of the same nature; and those that cause the other senses will be reducible to something of this kind, though we cannot explain it as distinctly.[28]

86

Doubtless, an infinite intellect would know to count each beat, measuring each vibration in relation to each other in an act of manifest arithmetic. Our finite, hearing minds content themselves with a partial and occult calculation. Without knowing the reasons for our pleasure, we thus come to sense it in an act of representation that Leibniz considered to be simultaneously clear and confused.[29] "When I recognize one thing among others without being able to say in what its differences or properties consist," he wrote in the *Discourse on Metaphysics* of 1686, "the recognition is confused. This is how we know sometimes clearly, without having any doubts, that a poem, a painting is well or poorly made, since there is in it a *je ne sçay quoy* that satisfies or shocks us."[30]

A few decades after Leibniz's death, Alexander Gottlieb Baumgarten gave a name to cogitations that share the characteristics of being both clear and confused. First in his *Philosophical Meditations Pertaining to Poetry* of 1735 and then in his massive treatise of 1750, he proposed that such perceptions be termed "aesthetic." These representations, he would explain, are clear, for they are distinct from others; but they are also confused, since we cannot reduce them to their elementary traits.[31] The use of the attribute "aesthetic" for such cogitations in itself involved a novelty. Before Baumgarten, the term had largely been employed for sensation in general. Henceforth it would increasingly evoke the perception of the beautiful and sublime. Yet the term pointed to a specific mode of representation that Leibniz had defined. Had the inventor of the monadology spoken in the idiom of his successors, therefore, he might also have argued that aesthetic pleasure is rational in kind and that it bears witness to an order that can be fully measured, albeit not by us. From the simultaneously clear and obscure sensation of a pleasure impenetrable to our senses, an inference, for Leibniz, could be drawn: a hidden harmony must obtain.

Were one to find the sights and sounds of this world to be aesthetic, then, one might well deduce that all of nature conceals an arrangement intelligible in itself. Being the creation of an infinite intellect, this world must be the best of all ordered totalities, arranged, in its parts and in its structure, to accord with a reasoned plan. It must be the best of all concordant wholes, with a maximum of pleasing commensurability and a minimum of incommensurability, also appreciable in itself. It is difficult to imagine the author of the *Theodicy* hesitating to argue such points. Yet can one be certain that the inference from pleasure to order is sound? Can one know, from seeing and listening, that the harmony of nature is truly real? After Benedetti, Beeckman, Leibniz, and Baumgarten, Kant would pose this question, recalling and rewriting the old problem of the harmonic transcription of the world. Submitting the idea of a beautiful and obscurely intelligible nature to a new critique, he would inevitably recall the one troubling possibility that the moderns, like the ancients and medievals long before them, could never altogether set aside: the possibility that despite the best of reasoned temperaments, nature might admit no ciphers, being neither commensurable nor incommensurable but simply unintelligible to us.

CHAPTER SEVEN

Of Measureless Magnitude

Kant maintained that the real history of philosophy began with Pythagoras. That claim, in its form, was doubtless unusual, but it was not absolutely novel. One might even consider it to be classical, for an ancient tradition, beginning with Heraclides Ponticus, holds that Pythagoras was the first thinker to claim for himself the name of "philosopher."[1] Yet Kant attributed a new meaning to this fact. In the "Short Outline of a History of Philosophy" contained in his *Lectures on Logic*, Kant declared that although "there is some difficulty in determining the borderline at which the *common* use of the understanding ends and the *speculative* begins, or where the common cognition of reason becomes philosophy," in principle, one may still distinguish "cognition of the general *in abstracto*" from "cognition of the general *in concreto*." One may dub the first variety of knowledge "speculative cognition," and one may call the second "common cognition." Then it can be argued that "among all peoples, the Greeks first began to philosophize, for they first attempted to cultivate the cognition of reason *in abstracto* without the guiding thread of pictures, while other peoples sought instead to make concepts intelligible to themselves *in concreto* by pictures only."[2] Nonetheless, Kant adds, the earliest thinkers to cultivate the speculative use of cognition did so but imperfectly. The Ionians and the Eleatics may have distinguished themselves in that first moment of speculative cognition, but they "clothed everything in pictures."[3] With their first propositions, philosophy had arisen. It had yet to attain, however, the purity it announced.

That step was soon taken. "Around the time of the Ionian school," Kant relates, "there arose in Greece a man of rare genius, who not only established a school but at the same time designed and carried out a project the like of which had never been before. This was Pythagoras, born on the isle of Samos."[4] In one of his last essays, "On a Newly Arisen Superior Tone in Philosophy," Kant once again recalled the unique importance of this archaic master. Here he presented Pythagoras as marking the very earliest point in the history of "the name of philosophy," understood as a synonym for "the scientific wisdom of life."[5] Recalling that philosophy did not begin with Socrates or Plato, Kant admonishes: "We must not forget *Pythagoras*." "Of course," he concedes, "too little about him is known to have a secure grasp on the metaphysical principles of his philosophy,"[6] but one may infer that Pythagoras sought the abstract principles of knowledge in general. One may also surmise that he took them to be numbers. Finally, by all accounts, Pythagoras discovered such quantities in sounds. Harmonic intervals, he showed, could be seen to express arithmetical ratios, which, in turn, could be detected both above the earth and within the soul. "History," Kant writes,

> says that the discovery of numerical relations between the tones and the precise law according to which they alone could be made into music brought Pythagoras to the following thought: because in the play of sensations mathematics (as the science of numbers) also contains the formal principles of these sensations (and indeed, as it appears, does so a priori, on account of its necessity), an albeit only obscure intuition dwells within us, and this intuition is of a nature that has been ordered according to numerical equivalents by an understanding whose rule extends over nature itself; once applied to the heavenly bodies, this idea can then produce the theory of the harmony of the spheres. Now, nothing is more enlivening to the senses than music; the enlivening principle in man is, however, the *soul*; and since music, according to Pythagoras, merely rests on perceived numerical relations and (something to note well) the enlivening principle in a human being—the soul—is at the same time a free, self-determining being, his definition

of the soul—*anima est numerus se ipsum movens* [the soul is a number that moves itself]—can perhaps be understood and to a certain extent even justified, if one assumes that the capacity to move on one's own was supposed to indicate the difference between the soul and matter, which in itself is lifeless and can be moved only by something exterior, and so this capacity was supposed to indicate freedom.[7]

By attending to tones and "the precise law according to which they could be made into music," Pythagoras would thus have reached a properly speculative cognition. He would have cast all "pictures" aside, setting numbers in their place. And with his obscure "intuition of a nature that has been ordered according to numerical equivalents," a first philosophy worthy of the name would have begun. By 1796, the year of his essay "On A Newly Arisen Superior Tone in Philosophy," Kant did not hesitate to identify how numbers could come to play such a role. Mathematical quantities furnished Pythagoras with a means to cognize as diverse a set of things as sensations, natural phenomena, heavenly bodies, and the soul, as he explains, because such arithmetical quantities were seen to display "purposiveness" (*Zweckmässigkeit*) in their form. Numbers, in other words, exhibited "a fitness to solve a manifold of problems, or a manifold of solutions to one and the same problem... starting from a single principle."[8] Just as Plato later proceeded to derive his "archetypes" or "Ideas" of things from a set of purposive (*zweckmässig*) geometrical figures, so Pythagoras reached his general cognition of nature by trusting in the "fittingness" (*Tauglichkeit*) of arithmetical multitudes.

But so, too, the archaic thinker ultimately allowed himself to be led astray. Like many after him, he mistook a purely mathematical order for the absolute. Pythagoras, the first real philosopher, thus became the first deluded visionary, falling prey to the danger that, for Kant, threatened all "speculative cognition." Believing himself to have "hit upon a secret" in discovering the laws of numbers, "and even believing himself to see something extravagantly great where he saw nothing," Pythagoras slid from thinking into feeling, and from the certainty of knowledge, he soared into the confused heights of "exaltation" (*Schwärmerei*).[9] Although he doubted that one might reconstruct Pythagoras's metaphysics with any precision, Kant, for his part, was convinced of the reality of that error. Without any trace of hesitation, the modern thinker recounted how, as his illustrious predecessor advanced in his investigations of the harmony of nature,

> his attention was attracted to the appearance of a certain purposiveness and an, as it were, intentionally implanted fitness in the character of numbers to solve numerous rational tasks of mathematics, where intuition a priori (space and time) and not merely discursive thought must be presupposed; he assumed a kind of *magic* simply in order to make comprehensible the possibility of extending not only our concepts of quantity in general but also their particular and, as it were, secret properties.[10]

Kant's own thought might be described as an attempt to hold fast to the activity of philosophy without giving way to any such "exaltation": to make full use of the concepts of the understanding without extending them beyond the field of their legitimate application and passing into a "kind of magic." In this sense, the task Kant set himself was to follow Pythagoras, yet only to a point. He would reach a general and abstract cognition without straying further.

It was to determine the possibilities and limits of the human faculties of cognition that Kant undertook his massive project of a "critique of pure reason." "Critique" was to establish the conditions and bounds of the legitimate use of "pure reason," defined as the ability to judge according to principles that are necessary, universally valid, and independent of experience—in short, "a priori." It was an axiom of Kant's critical system that the power of judgment must employ two types of concepts, which determine two different kinds of objects. These are, on the one hand, the concepts of nature and, on the other, the concept of freedom. For Kant, this division of the concepts of pure reason dictates a partition of the "domain" (*Gebiet*, *ditio*) of philosophy into the "theoretical" and the "practical."[11] Kant's third and last treatise on pure reason, the *Critique of Judgment* of 1790, offers the most important formulation of the doctrine of this partition. "Concepts of nature make possible a *theoretical* cognition governed by a priori principles," Kant writes,

> whereas the concept of freedom carries with it, as far as nature is concerned, only a negative principle (namely, of mere opposition), but gives rise to expansive principles for the determination of the will, which are therefore called practical; hence we are right to divide philosophy into two parts that are quite different in their principles: theoretical or *natural philosophy*, and practical or *moral philosophy* (*morality* is the term we use for reason's practical legislation governed by the concept of freedom).[12]

The distinction between the two types of concepts can be simply stated. Each variety has its correct use, or, as Kant writes, its proper "legislative" power. The concepts of nature may be referred to an object insofar as it is defined by the twin transcendental conditions of intuition, namely, space and time. For this reason, the natural or theoretical domain extends to appearances, and to them alone. The concept of freedom, by contrast, refers to an object that cannot be grasped according to the forms of intuition. This is a thing in itself: the unconditioned freedom of the will, which constitutes the sole ground of moral and lawful action. Kant also relates the two types of concept to two faculties of the single power of pure reason. "Legislation through concepts of nature," Kant specifies, "is performed by the understanding (*Verstand*) and is theoretical. Legislation through the concept of freedom is performed by reason (*Vernunft*) and is merely practical."[13] To two types of concepts, those of nature and that of freedom, there correspond, therefore, two types of objects, the conditioned and the unconditioned; two domains of philosophy, the theoretical and the practical; and two different powers of the mind, the understanding and reason. These divisions are clear and, on the surface, at least, they are unbridgeable. "The concept of nature does indeed allow us to present its objects in intuition, but as mere appearances rather than as things in themselves, whereas the concept of freedom does indeed allow us to present its object as a thing in itself, but not in intuition."[14] The domain of nature, therefore, may be apprehended by cognition as it appears, but never as it is in itself, and, with perfect symmetry, the domain of freedom may be conceived by thought as it is in itself, yet not as it appears. The two types of notions share, therefore, one fundamental trait: neither the concepts of nature nor the concept of freedom "can provide us with theoretical cognition of its objects...as things in themselves."[15]

Nonetheless, Kant maintains that the conditioned and the unconditioned, appearances and the thing in itself, nature and freedom, must be conceived as each implying a substrate, to which he gives a single name: the "supersensible." That the concept of freedom presupposes an intelligible being is surely manifest: the unconditioned will is nothing if not withdrawn from the order of phenomena, and, insofar as it is "unconditioned," it is by definition supersensible in nature. Yet appearances, too, Kant explains, imply a nonsensuous stratum, precisely insofar as they are the appearances of things in themselves unknowable to us. "We do need the idea of the supersensible in order to base on it the possibility of those objects of experience," Kant affirms, while adding: "But the idea itself can never be raised up and expanded into a cognition."[16] If one considers the status of such an "idea," one will conclude that it constitutes not a "domain," but a "realm" (Feld), which forms a ground common to the otherwise disparate concepts of nature and freedom. No cognition, Kant argues, may be erected in this shared "supersensible realm." But for the needs of thought, it can nonetheless legitimately be occupied:

> There is a realm that is unbounded, but that is also inaccessible to our entire cognitive power: the realm of the supersensible. In this realm we cannot find for ourselves a territory on which to set up a domain of theoretical cognition, whether for the concepts of the understanding or for those of reason. It is a realm that we must indeed occupy with ideas that will assist us in both the theoretical and the practical use of reason; but the only reality we can provide for these ideas, by reference to the laws [arising] from the concept of freedom, is practical reality, which

consequently does not in the least expand our theoretical cognition to the supersensible.[17]

While this unbounded realm is fundamentally "inaccessible to our entire cognitive power," the idea of it plays a crucial role in Kant's system of pure reason. It resolves a major question in the critical doctrine of the freedom of the will. Kant evokes the problem and the solution in a single, curious modal syllogism. We pass from an "is" to an "ought" before reaching the certitude of a "must." In a first step, Kant underlines the abyss separating the sensible from the intelligible: "An immense gulf is fixed between the domain of the concept of nature, the sensible, and the domain of the concept of freedom, the supersensible, so that no transition from the sensible to the supersensible... is possible, just as if they were two different worlds, the first of which cannot have any influence on the second." In a second step, Kant recalls that if the free will cannot render itself effective in nature, it will be vain. Thus he formulates this practical imperative: "And yet the second *ought* to have an influence on the first, i.e., the concept of freedom is to actualize in the world of sense the purpose enjoyed by its laws." Now Kant reaches his conclusion. The unbridgeable "gulf" must be bridged:[18] "Hence it must be possible to think of nature as being such that the lawfulness in its form will harmonize with at least the possibility of [achieving] the purposes that we are to achieve in nature according to the laws of freedom. So there must after all be a basis *uniting* the supersensible that underlies nature and the supersensible that the concept of freedom contains practically."[19]

The "basis" reached by Kant in these lines stretches across the "immense gulf" that he has stipulated must lie between the domains of nature and freedom. Beneath "the supersensible that underlies nature and the supersensible that the concept of freedom contains practically," it constitutes a foundation laid in the site of a structural abyss, a ground that must occupy a place of groundlessness. Kant hardly conceals why the critical philosophy must posit such a paradoxical expanse in the exact delimitation of the territory and domains of its jurisdiction. This "basis" alone, he affirms, will make it "possible to think of nature as being such that the lawfulness in its form will harmonize with at least the possibility of [achieving] the purposes that we are to achieve in nature according to the laws of freedom." Without the positing of such a "supersensible realm," in short, the domains of nature and freedom would be irreducibly at odds: it would be inconceivable how the intelligible will might ever realize itself within the world of appearances, and the theoretical and practical branches of philosophy, as a consequence, would be utterly disparate. Only through the positing of such a ground may the domains of nature and freedom be made to "harmonize" (*übereinstimmen*), and only then may the critical philosophy Kant strives to elaborate be rendered a coherent whole.

In the *Critique of the Faculty of Judgment*, however, Kant does more than demand the accord of sensible nature with the intelligible principle of the will. He also identifies a specific power of cognition that bears witness to its reality. This principle is that of a "reflective judgment." In the preface to this work, Kant explains why, after having completed a *Critique of Pure Reason* and a *Critique of Practical Reason*, he found it necessary to propose a Third *Critique*:

> It was actually the *understanding*, which has its own domain as a *cognitive power* insofar as it contains principles of cognition that are constitutive a priori, which the critique that we call the critique of pure reason was to make the secure and sole possessor [of that domain] against all other competitors. Similarly *reason*, which does not contain any constitutive a priori principles except [those] for the *power of desire*, was given possession [of its domain] by the critique of practical reason.[20]

Kant has delimited the domain of nature, which corresponds to the understanding and the constitutive concepts of pure reason; he has drawn the borders of the domain of freedom, which corresponds to the faculty of desire and regulative principles of practical reason. Yet a critique "of our ability to judge according to a priori principles," we read, "would be incomplete if it failed to include, as a special part, a treatment of judgment; judgment must be treated, in a special part of the critique, even if in a system of pure philosophy, its principles are not such that they can form a special part between the theoretical and practical philosophy, but may be annexed to one or the other as needed."[21]

This statement may well have startled the readers of Kant's earlier works. Not only does the philosopher now affirm that the critique of pure reason must admit, as a "special part," a treatment of judgment, whose principles can, by nature, form no "special part between the theoretical and practical philosophy." There is more than the positing of this landless "part," "to be annexed as needed." No less unexpected is the express indication that an autonomous "treatment of judgment" must be appended to the First and Second *Critiques*. In appearance, at least, those two treatises were themselves nothing if not accounts of judgment. The *Critique of Pure Reason* had investigated the possible grounds of a priori synthetic judgments; the *Critique of Practical Reason* had determined the legitimate application of the categories of the moral law to individual cases. But in publishing the *Critique of Judgment*, Kant disclosed that his earlier works had concerned themselves with but one of two distinct forms of judgment. The earlier treatises had both implicitly assumed that all judgment is "determinative" (*bestimmend*), in that it consists of subsuming a particular under a universal principle, such as a rule or concept, which is given in advance. Yet judgment may also be "reflective" (*reflektierend*) in structure: in such cases, "only the particular is given, and judgment has to find the universal for it."[22] Both varieties of judgment, we learn, operate by a priori principles. In determinative judgment, the faculty of the understanding may furnish the rules by which we subsume particulars under the universal transcendental concepts of nature. Kant now admits that there are cases in which such a process is inadequate: "since the laws that pure understanding gives concern only the possibility of a nature as such (as an object of sense), there are such diverse forms of nature, so many modifications, as it were of the universal transcendental concepts of nature, which are left undetermined by these laws, that surely there must be forms for these laws too."[23]

Confronted with the diversity of nature, the mind takes recourse to "reflection," lingering with the given particular, even as it searches for the general rule that will apply to it. "Hence reflective judgment," Kant argues, "which is obliged to ascend from the particular in nature to the universal, requires a principle, which it cannot borrow from experience, precisely because it is to be the basis for the unity of all empirical principles under higher, though still empirical principles, and hence is to be the basis that makes it possible to subordinate empirical principles to one another in a systematic way."[24] The transcendental principle of this form of judgment can be simply named: it is that by which one may view things with regard to the intended reason for their actuality—in a word, their "purpose." "Insofar as the concept of an object also contains the basis for the object's actuality, the concept is called the thing's *purpose*, and a thing's harmony with that character of things which is possible only through purposes is called the *purposiveness* [*Zweckmässigkeit*] of its form."[25] This definition of "subjective" or "purely formal" purposiveness constitutes one of the major novelties of the *Critique of Judgment*, in which it plays a systematic role.[26] In addition to the faculties of the understanding and reason, the philosopher now admits a power of "reflective judgment" whose task is to mediate between them. In addition to the principle of lawfulness, familiar to the understanding, and the principle of final purpose, familiar to reason, he now introduces the principle of purely formal "purposiveness." And in addition to the domains of nature and freedom, on which the concepts of the understanding and reason may bear, Kant posits a third realm, in which reflective judgment will apply: that of the beautiful and the sublime.

Yet there is more. Having discovered the principle of reflective judgment, Kant uncovers a "mental power" (*Erkenntnisvermögen*) of considerable novelty. He had tied the faculty of understanding to the power of cognition, even as he had linked the faculty of reason to the power of desire. Now he relates the faculty of judgment to a power of an apparently different nature: "feeling" (*Gefühl*) and, more precisely, "the feeling of pleasure and displeasure."[27] This is a striking sentiment. It cannot be conceived as the result of any sensible modification of the judging subject, such as an empirical attraction or repulsion to an object, for then it would be empirical rather than transcendental in kind. Nor can it be defined as resulting from a rational inclination, such as a moral interest, for then it would pertain to the faculty of reason rather than to judgment. Kant argues that the feeling of pleasure and displeasure is the immediate and necessary correlate of the perception of purely formal purposiveness in nature. "It is a fact," the thinker writes, "that when we discover that two or more heterogeneous empirical laws of nature can be unified under one principle that comprises them both, the discovery does give rise to a quite noticeable pleasure, frequently even admiration, even an admiration that does not cease when we have become fairly familiar with the object."[28] The apprehension of the suitability of natural phenomena to the cognitive faculties is pleasing in itself. "By contrast," Kant continues,

> we would certainly dislike it if nature were presented in a way that told us in advance that if we investigated nature slightly beyond the commonest experience we would find its laws so heterogeneous that our understanding could not unify nature's particular laws under universal empirical laws. For this would conflict with the principle of nature's subjectively purposive specification in its genera, and with the principle that our reflective judgment follows in dealing with nature.[29]

Unlike Leibniz, Kant does not suggest that any knowledge may be drawn from the pleasing apprehension of the harmonious arrangement of natural kinds. On the contrary, it is one of his major and most famous theses that reflective judgments imply no determinate claims to cognition. Kant writes of "*subjective* purposiveness" precisely to distinguish this a priori principle from the principle of "objective purposiveness," which always presupposes a determinate concept of a purpose, whether in kind external ("utility") or intrinsic ("perfection").[30] Subjective purposiveness operates independently of such determinate concepts, being, in Kant's famous phrase, "purposiveness without purpose" (*Zweckmässigkeit ohne Zweck*).[31] A judgment of subjective finality "brings to our notice no characteristic of the object, but only the purposive form in the [way] the presentational powers are determined in their engagement with the object."[32] Insofar as it may be referred to what is "merely subjective in the presentation of an object," the reflective judgment of purposiveness may be termed "aesthetic" in character.[33] Kant concedes that much of this character may, in part, enter into knowledge. Nonetheless, he argues that the purposiveness of a presentation immediately produces in the judging subject a feeling distinct from cognition. "For through this pleasure or displeasure," Kant explains, "I do not cognize anything in the object of the presentation."[34] Yet it would be an error to conclude that the feeling tied to purposiveness, therefore, is empirical because subjective. This sentiment of pleasure and displeasure also lays claim to universality, because it testifies to a principle that, being a priori, is common to all minds:

> When the form of an object (rather what is material in its presentation, viz., in sensation) is judged in mere reflection on it...to be the basis of a pleasure in such an object's presentation, then the presentation of this object is also judged to be connected necessarily with this pleasure, and hence connected with it not merely for the subject apprehending this form but in general for everyone who judges it. The object is then called beautiful, and our ability to judge by such a pleasure (and hence also with universal validity) is called taste.[35]

On several major levels, this theory of judgment is a doctrine of harmony. Kant describes the purposive suitability of nature to our faculties as a "correspondence" (*Übereinstimmung*), "accord" (*Einstimmung*), and "harmony" (*Harmonie*), as if these words were, in essence, synonymous with each other.[36] He does not hesitate to equate the character of being "subjectively purposive" with the quality of being simply "harmonious" (*harmonisch*).[37] In the terms of the *Critique of Judgment*, to deem an object aesthetically pleasing is to consider it as "harmonizing" with the power of pure reason, and to judge nature as beautiful is to view it as being, as it were, in accord with the vocation of man. Only then—in the perception of a sensible world somehow in tune with the powers of human cognition—may one find an indication of that "supersensible realm" conceived and indeed demanded by the opening of the Third *Critique*.

Yet a second dimension of harmony may also be discerned in the pages of this work. The subjective correlate of the "correspondence" of nature with our faculty is, in turn, an "attunement" (*Stimmung*) of the faculties themselves: if I judge an object to be beautiful, it can be only because its form has caused the elements of my cognitive power to enter into a pleasing play of accords among themselves. The "subjective universal communicability" contained in a judgment on beauty, Kant writes, "can be nothing but [that of] the mental state in which we are when imagination and understanding are in free play (insofar as they harmonize with each other [*sofern sie unter einander . . . zusammen stimmen*])."[38] The faculties of concepts—the understanding and reason—are then said to "accord" themselves with the faculty of exhibition, that is, the imagination; to become aware of beauty is thus "to become conscious . . . of a reciprocal subjective *harmony* between the cognitive powers [*Übereinstimmung der Erkenntniskräfte unter einander*]."[39] Finally, one might also discern a third harmony in the theory of taste "as a kind of *sensus communis*," to which Kant dedicates a crucial section of the Third *Critique*.[40] Aesthetic judgments, we learn, are by their nature universally "communicable" (*mitteilbar*), in the sense that the expression of the attunement of one faculty immediately demands the accord of all others.[41]

One might therefore expect Kant's treatise on aesthetic judgment to contain a major analysis of harmony in the traditional sense of musical order. But in the *Critique of Judgment*, music plays a relatively minor role. Here, too, the contrast with Leibniz is striking. One may presume that Kant was well aware of his predecessor's doctrine of music as an "unconscious arithmetic." Kant himself also took a special interest in the modern science of vibratory bodies, as several remarks in his treatise indicate. In a chapter devoted to the elucidation of the theory of taste "by examples," he raises the question of "tone." Taking recourse to recent developments in the physics of light and sound, Kant aims to settle a question that troubled his transcendental analysis of judgment: the question, namely, whether the tones of color, like those of music, may be considered beautiful in themselves. Referring to Leonhard Euler, the great mathematician, physicist, and theorist of temperament, Kant writes:

> If, following *Euler*, we assume that colors are vibrations (*pulsus*) of the aether in uniform temporal sequence, as, in the case of sound, tones are such vibrations of the air, and if we assume—what is most important (and of this, after all, I do not doubt at all)—that the mind perceives not only, by sense, the effects that these vibrations have on the excitement of the organ, but also, by reflection, the regular play of impressions (and hence the form in the connection of different presentations), then color and tone would not be mere sensations but would already be the formal determination of the manifold in these, in which case they could even by themselves be considered beauties.[42]

The statement has the allure of both clarity and certainty. If it may be granted that colors, like tones, result from a series of periodic vibrations per instant, then even in their simplest forms, these sensible qualities will constitute more than material; they will be, as Kant writes, "formal determinations of the manifold," and as such, they will be judged subjectively purposeful "beauties" in themselves, without consideration of their many possible forms of arrangement. But the textual history of this passage betrays the signs of some hesitation. In the first edition of the *Critique of Judgment*, Kant had cited Euler's same findings, remarking not "of this, I do not doubt at all" (*woran ich gar nicht zweifle*), as in the second edition, but rather: "of this, I very much doubt" (*woran ich doch gar sehr zweifle*).[43] The question is far from trivial, as several commentators have observed.[44] If Euler is correct, then visual and acoustical tones may be "beautiful" (*schön*) in themselves. If he is wrong, such qualities may be beautiful only in their arrangement, being on their own merely "agreeable" (*angenehm*), like all other presentations reducible to mere sensible material.[45] Although Kant does not state it openly, it seems the question of the precise status of tone—beautiful or agreeable, formal or material—must be decided for critical philosophy by the sciences of nature.

At the same time, however, Kant expressly insists that the physical and mathematical structure of harmonic sounds may play no role in their judgment. In the one section of the Third *Critique* devoted in part to the art of music, Kant thus declares, without hesitation:

> Mathematics certainly do not play the slightest part in the charm and mental agitation that music produces. Rather, they are only the indispensable condition (*conditio sine qua non*) of that ratio of the impressions, in their combination as well as change, which enables us to comprehend them; and thus impressions are kept from destroying one another, so that they harmonize in such a way as to produce, by means of affects consonant with [this ratio], a continuous agitation and quickening of the mind, and thus they produce an appealing self-enjoyment.[46]

Almost every major term in this statement is susceptible to a double reading, for in the very moment in which Kant sets aside the mathematical regularities that "certainly do not play any part in the charm and mental agitation that music produces," he appropriates for his theory a set of expressions drawn from the old art of harmonic sounds. "Ratio" (*Proportion*), "harmonization" (*zusammenstimmen*), and "consonance" (*konsonieren*) are called upon to displace the discourse from which they derive. One might well wonder, therefore, whether Kant here truly rejects the old paradigm of a harmony intelligible by mathematical means, as he suggests. He could also be said to raise it to a new level of generality, where its terms come to define aesthetic pleasure as such.

"Mathematics," in any case, does play a part in Kant's doctrine of judgment. Quantities, defined in commensurate and incommensurate relations, become central problems at one crucial juncture in the *Critique of Judgment*. At this point, Kant seeks to consider the aesthetic representations of the mind not in their proportions, but in their disproportions, and to define not the purposiveness of nature, but rather its "contra-purposiveness." This point is the Analytic of the Sublime. Kant presents this brief section of his book as "a mere appendix" (*ein bloßer Anhang*) to the Analytic of the Beautiful.[47] In the "Transition from the Power of Judging the Beautiful to That of the Sublime," he indicates that in several respects, the second discussion is a pendant to the first. The judgment of the sublime, like that of the beautiful, is reflective in structure. As Kant reminds the reader, it must therefore bear on a singular object, which, when referred to the faculty of cognition as a whole, provokes a feeling that depends neither on sensory material, as does the agreeable, nor on a determinate concept, as does the good.[48] "Hence," Kant reasons, "the liking is connected with the mere exhibition or power of exhibition, i.e., the imagination, with the result that we regard this power, when an intuition is given us, as harmonizing with the *power of concepts*, i.e., the understanding or reason, this harmony furthering the aims of these."[49] For this reason, judgments of the sublime lay claim to the same universality as do judgments of the beautiful: both "proclaim themselves valid for all subjects, though what they lay claim to is merely the feeling of pleasure, and not any cognition of the object."[50]

Kant proceeds, however, to distinguish the sublime from the beautiful in several respects. The judgment of beauty concerns the form of the judged object, which is perceived as harmonizing with the powers of the mind. "But the sublime," we read, "can also be found in a formless object, insofar as we present *unboundedness*, either [as] in the object or because the object prompts us to present it, while yet we add to this unboundedness the thought of its totality."[51] For this reason, the philosopher argues that the sublime sets in play an accord of the imagination not with the understanding, as in the beautiful, but with reason, totality being a concept whose application can be only practical.[52] In the beautiful, moreover, "our liking is connected with a *quality*"; in the sublime, it is bound to the "presentation of *quantity*."[53] A further trait opposes the two kinds of "liking." The enjoyment of the beautiful follows from the apprehension of finality; it is, therefore, immediate in nature, "carrying directly with it a feeling of life's being furthered." The enjoyment of the sublime, by contrast, is indirect: "it is produced by the feeling of a momentary inhibition of the vital forces followed immediately by an outpouring of them that is all the stronger."[54] One may thus describe the affect that it engenders as a "negative pleasure" (*negativer Lust*).

"But the intrinsic and most important distinction between the sublime and the beautiful," Kant writes, "is the following": the judgment of beauty concerns what appears commensurate with our faculties, the sublime, what seems incommensurate with us. "If something arouses in us, merely in apprehension and without any reasoning on our part, a feeling of the sublime, then it may indeed appear, in its form, contra-purposive [*zweckwidrig*] for our power of judgment, incommensurate with our power of exhibition [*unangemessen unserm Darstellungsvermögen*], and as it were violent to our imagination, and yet we judge it all the more sublime for that."[55] For this reason, the sublime, unlike the beautiful, never promises to "expand our cognition of natural objects." "In what we usually call sublime in nature," Kant writes, "there is such an utter lack of anything leading to particular objective principles and to forms of nature conforming to them, that it is rather in its chaos that nature most arouses our ideas of the sublime or in its wildest and most rule-less disarray and devastation."[56] The aesthetic value of such nature lies in the use we make of it. As Kant aims to show, the mind can recover a purposiveness beyond—and indeed against—the "contra-purposiveness" perceived in the sensible world. Then it may find harmony even "in its wildest and most ruleless" disharmony.

Perhaps nowhere in his doctrine of judgment does Kant confront a more radical disproportion than in his consideration of the "mathematical sublime." Something worthy of that designation, Kant writes, "is *absolutely* large (*schlechthin groß*)."[57] Such a quantity should not be confused with the determination of a certain size, for "saying simply (*simpliciter*) that something is large is quite different from saying that it is *absolutely large* (*absolute, non comparative magnum*). The latter is *what is large beyond all comparison*."[58] Kant distinguishes, in this sense, between being a "magnitude" and possessing a certain "size." To know that something is a magnitude (*quantum*,

Größe) "may be cognized from the thing itself, without any comparison to others," but to know that something possesses a particular size (*magnitudo*), we must employ a "measure" (*Maß*).[59] One defines *magnitudo* by a measure, whose unit may be arbitrarily assigned. By nature, such determinations of quantity will always yield relative sizes; given any definite mass, one may imagine a greater or lesser one. The measurement of physical beings, Kant writes, illustrates this fact: telescopes and microscopes have amply shown that "nothing in nature can be given, however large we may judge it, that could not, when considered in a different relation, be degraded all the way to the infinitely small, nor conversely anything so small that it could not, when compared with still smaller standards, be expanded for our imagination all the way to the magnitude of a world."[60] The "absolutely large," by contrast, necessarily outstrips all standards of comparison. Incommensurate with every given *quantum*, its magnitude is measureless.

One might suppose that such a quantity would find no place in a treatise on aesthetic judgment, defined as the apprehension of presentations in intuition. Kant, indeed, expressly states that "the sublime must not be sought in things of nature, but must be sought solely in our ideas."[61] But in the Analytic of the Sublime, he seeks to demonstrate this unlikely fact: the thought of the incomparably large may be incited by the perception of the very sensible bodies that lack it. Kant evokes several possible cases. The judging mind may be faced with a technical artifact of impressive proportions, such as the pyramids of Egypt or St. Peter's Basilica in Rome.[62] But the contemplation of mere landscapes may also suffice. "Nature offers examples of the mathematically sublime, in mere intuition, whenever our imagination is given, not so much a larger numerical concept, as a large unity for a measure."[63] If, for instance, one takes as a standard "a tree that we estimate by a man's height," seeking to conceive, by means of it, the dimensions of a mountain, the imagination will find that it cannot present to itself the entirety of the being with which it is confronted. Halting, it will be remitted to the bounds that define its own activity.

Kant explains that whenever the imagination seeks to determine the relative size of a certain magnitude, it must perform two activities: "apprehension (*apprehensio*)" and "comprehension (*comprehensio aesthetica*"). Apprehension, which proceeds part by part, may continue *ad infinitum*; there is no reason why it must reach any limit. Not so the aesthetic faculty of presenting parts in unity:

> Comprehension becomes more and more difficult the farther apprehension progresses, and it soon reaches its maximum, namely, the aesthetically largest basic measure for an estimation of magnitude. For when the apprehension has reached the point where the partial presentations of sensible intuition that were first apprehended are already beginning to be extinguished in the imagination, as it proceeds to apprehend further ones, the imagination then loses as much on one side as it gains on the other; and so there is a maximum in comprehension that it cannot exceed.[64]

At that upper limit, the imagination "strives to exceed" itself, and it fails, "sinking back into itself."[65] Aesthetic estimation, Kant writes, now finds itself "pushed to the point where the ability of our imagination is inadequate to exhibit the concept of magnitude."[66]

In this sublime play, however, one faculty, faltering, summons another. The moral power of desire intervenes. Even as the imagination fails to render the series of many parts in a single whole, "reason demands comprehension in *one* intuition, and *exhibition* of all the members of a progressively increasing numerical series, and it exempts from this demand not even the infinite (space and time). Rather, reason makes us unavoidably think of the infinite (in common reason's judgment) as *given in its entirety* (in its totality)."[67] In the moment the imagination balks before the relative enormity of a single physical *quantum*, reason commands the judging mind to ponder a mathematical quantity of another nature, to which no standard, be it sensible or mathematical, may be adequate: "the infinite." For Kant, this is a rational magnitude, which is strictly measureless. That it necessarily exceeds "the world of sense" follows from its definition as absolute in proportion. Now he adds that it also exceeds the power of cognition: even "mathematical estimation by means of numerical concepts" cannot render the infinite, "given in its entirety," which is to say, conceived "in its totality."[68] Only the moral faculty may respond to the demand to "think of the infinite," furnishing an idea of reason for what can be neither perceived nor known.

As it accedes to a positive infinity through the "voice of reason," the mind comes to feel this moral fact: it has "within itself a power that is supersensible, whose idea of a noumenon cannot be intuited but can yet be regarded as the substrate underlying what is mere appearance, namely, our intuition of the world."[69] Confronted with the disharmony of "apprehension" and "comprehension," the faculty of judging becomes aware of a far more fundamental incommensurability, which distinguishes all appearance from the thing in itself, nature from freedom. The mind now ponders the "supersensible substrate" that, for the sake of the realization of the will in nature, phenomenon and noumenon both presuppose:

> That magnitude of a natural object to which the imagination fruitlessly applies its entire ability to comprehend must lead the concept of nature to a supersensible substrate (which underlies both nature and our ability to think), a substrate that is large beyond any standard of sense and hence makes us judge as *sublime* not so much the object as the mental attunement in which we find ourselves when we estimate the object.[70]

In this passage from the failure of the imagination to the emergence of the voice of reason, an initial sentiment of inadequacy thus gives way to "respect": moral admiration for the "superiority of the rational vocation of our cognitive powers over the greatest power of sensibility."[71] Now the mind accedes to the specific quality of "liking" connected to the judgment of the sublime, experiencing one "pleasure that is possible only by means of a displeasure."[72] The path is "indirect," as Kant himself maintains, but its end is unambiguous. Through a profound mental "agitation" (*Bewegung*), which may be "compared with a vibration [or "trembling," *Erschütterung*], that is, with a rapid alternation of repulsion from, and attraction to, the same object," the cognitive power reaches a final tone of considerable elevation.[73] By "violent" means, the mathematical sublime ultimately remits the judging faculty to the pleasure of reflective judgment: the transcendental feeling of a harmony of the domains of the will and nature.

One might also argue that it returns the mind to the old "feeling" of Pythagoras. Had that "man of rare genius," by Kant's own account, not also intuited in natural things a certain "purposiveness," discernable "in the play of sensations" rendered intelligible by the formal principles of mathematics? Despite its novelties, Kant's mathematical analytic of measureless magnitudes bears an unlikely resemblance to the antique theory of measured multitudes. Some readers might go so far as to call one the spectral repetition of the other. Kant's critical project, however, carries the old paradigm of harmony to its limit. A pleasing order may be found, even in the absence of all tones, proportions, and commensurabilities. A "mathematical" attunement can be reached, even where there are not multitudes, but magnitudes, and where one magnitude, in particular, is measureless by nature. A "purposiveness," finally, can be glimpsed, not only in the absence of purpose, but even in the face of its exact negation. In the disorder of "contra-purposiveness," an order of a kind, in short, prevails, even if it can assure no knowledge.

"Philosophy," for Kant, could surely go no further. It could not venture forward or backward with respect to this threshold without ceasing to be itself. The two possibilities for its movement are all too easily imaginable. To step beyond the point of discerning in nature a merely "subjective purposiveness" would be to lend to the "obscure intuition" of an accord of the will with nature the very certainty that the *Critique of Pure Reason* had dispelled. This would be to pass from cognition into visionary "enthusiasm" and "exaltation." Yet to step back from the Kantian position, renouncing the promise of "subjective purposiveness," would be to abandon the harmony of nature and the project of philosophy practiced since Pythagoras. This might be to study the human cognitive power without linking it to a conception of the ultimate unity of nature; it might also be to investigate the physical world, in its total organization, without wondering about the confirmation that that organization might offer to the moral vocation of the will. Kant attained a subtle, yet firm, position, which enabled him to hold fast to both the ancient Pythagorean paradigm of harmony and the new critical rule that stipulated that no cognition be derived from the pleasures of "obscure intuition." To do so, however, Kant had to accept a single axiom. The question of the formal unity of nature was to be conceived in purely disjunctive terms: as ordered or disordered, commensurate or incommensurate. In short: as beautiful or sublime. Then, whether nature appeared purposive or "contra-purposive," pleasing or displeasing, equivalences could be established. Promising proportions might be defined, even in aesthetic disproportion.

There was, however, also another possibility, and Kant knew it well. What if nature were neither harmonious nor disharmonious, but unintelligible to us? What if it were to possess no single form, which our faculties might grasp and judge? In the introductions to the *Critique of Judgment*, Kant observed that his transcendental doctrine required that "experience be regarded [...] as a system and not as a mere aggregate."[74] That was a view not given, but demanded, for the sake of the coherence of the critical philosophy. "It does not follow from this," Kant wrote,

> that nature is, even in terms of its *empirical laws*, a system which the human cognitive power *can grasp* and that the thorough systematic coherence of its appearances is an experience, and hence experience itself as a system, is possible for human beings. For the empirical laws might be so diverse and heterogeneous that, though we might on occasion discover particular laws in terms of which we could connect some perceptions to [form] an experience, we could never bring these empirical laws themselves under a common principle [and so] to the unity [characteristic] of kinship. We would be unable to do this if—as is surely possible intrinsically (at least insofar as the understanding can tell a priori)—these laws, as well as the natural forms conforming to them, were infinitely diverse and heterogeneous and manifested

themselves to us as a crude chaotic aggregate without the slightest trace of a system.[75]

Were nature indeed "so diverse and heterogeneous," were it but a "crude chaotic aggregate without the slightest trace of a system," even the Kantian solution would be unfounded. Devoid of the last shadow of proportion that is disproportion, nature would be neither "purposive" nor "contra-purposive," neither pleasing nor displeasing in its form. It would possess no one form at all. A "crude, chaotic aggregate," it would be ungraspable, and "without the slightest trace of a system," it would, in its sheer heterogeneity, incite in us no feeling. It is worth considering the consequences that would follow were such an "intrinsic possibility" to be realized. "Cognition *in abstracto*," then, would meet its end. "Cognition *in concreto*," "thinking by pictures," and the "common use of the understanding" might persist. Yet in the world the philosopher here both called to mind and banished from his thought, the infinite diversity of laws would render every "obscure intuition" into the order of all nature essentially suspect. Some structure in the cosmos, to be sure, might surely once more be imagined. But in this universe, which the critical Pythagorean conceived in passing and in dread, harmony could never again be real.

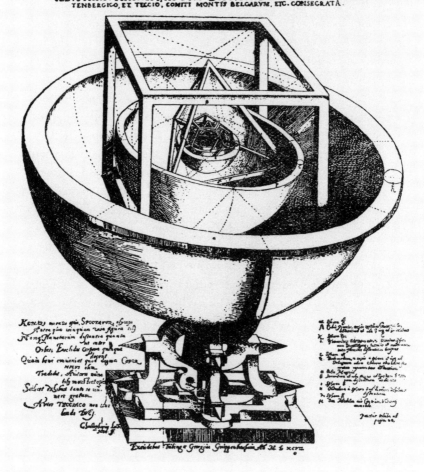

Nesting of Polyhedrons, *Mysterium Cosmographicum,* 2nd ed. (Frankfurt, 1621). (Photo: Dietmar Katz. bpk, Berlin/Art Resource, NY.)

CHAPTER EIGHT

Out of This World

Johannes Kepler was of the opinion that the true teachings of Pythagoras came to light long after the death of the fabled Greek philosopher. That claim might be understood to rest on a simple fact: the existing accounts of Pythagoras and his doctrines were all composed by disciples and commentators, successors who, while of different eras, lived long after the master of Samos. Yet Kepler's was a far less obvious point. In a certain sense, he maintained, the project initiated by Pythagoras began to reach completion not only after the archaic Greek period but also after antiquity as a whole and, indeed, even after the many centuries of the Middle Ages had come to a close. As Kepler once remarked in a letter to his beloved teacher, Michael Mästlin, there are reasons to hold that Pythagoras's project came to fruition in the age of modern science alone. One might well imagine that some two thousand years after the apparent death of the philosopher in an unknown Greek colony of southern Italy, Pythagoras's soul migrated into the body of a man born in late sixteenth-century Germany: more exactly, that of the astronomer, mathematician, metaphysician, and theologian born on December 27, 1571, in a city of moderate size in Württemberg and baptized Friedrich Johannes Kepler.[1] In his letter, the twenty-seven-year-old Kepler advanced this idea lightly, and doubtless it was at least partly in jest. Yet it is nonetheless worthy of consideration, and not only because many sources indicate that Pythagoras did espouse a theory of the transmigration of souls.[2] From his first works to his last, Kepler's reference to Pythagoras was constant. One may interpret this detail by noting the fact that the tradition attributes to the Pythagoreans the first account of the cosmos in which all celestial bodies, the Earth included, circle around the Sun, as Coperncius himself was well aware.[3] Yet the truth is that Kepler's allusions to Pythagoras expressed more than his allegiance to heliocentrism. They testified to the aim of Kepler's own philosophical and scientific project, which, while unmistakably bound to the past, was to lead to the most modern of discoveries. Like the Pythagoras he imagined long before him, Kepler aimed to grasp the proportions of the natural world by mathematical means. He sought to find, in the language of quantity, the fundamental elements that would allow him to transcribe the order of the universe.

Kepler already identified Pythagoras as his "guide, master and precursor" in the opening pages of his first great work, published in Tübingen in 1596. This was a slim volume to which the young astronomer gave the title *An Introduction to the Cosmographical Treatise, Containing the Cosmographical Mystery Concerning the Remarkable Proportion of the Heavenly Spheres, and Concerning the Genuine and Proper Causes of the Number, Magnitude, and Periodic Motions of the Spheres, Demonstrated by Means of the Five Regular Geometric Solids.* Today it is known more simply as the *Mysterium cosmographicum* or *The Secret of the Universe.*[4] "It is my intention, Reader," Kepler wrote in his preface, "to show in this little book that the most great and good Creator, in the creation of this moving universe, looked to those five regular solids, which have been so celebrated from the time of Pythagoras and Plato down to our own, and that He fitted to the nature of those solids, the number of the heavens, their proportions, and the law of their motions."[5] Kepler argued that to grasp the distances between the planets, it suffices to represent the relations that result from embedding the five regular convex polyhedra within each other; a simple set of proportions then emerges. Kepler demonstrated his thesis by inserting a cube between Saturn and Jupiter; a tetrahedron between Jupiter and Mars; a dodecahedron between Mars and Earth; an octahedron between Earth and Venus; and an icosahedron, finally, between Venus and Mercury.[6] Suddenly, Kepler maintained, the mathematical design of the universe comes to light. It may be seen that quantitative relations are inscribed in the order of the cosmos, and by necessity:

> "For it neither is, nor was right" (as Cicero says in his book on the universe, following Plato's *Timaeus*) "that the best should make anything except the most beautiful." Since, then, the Creator conceived the Idea of the universe in his mind...and it is the Idea of that which is prior, indeed, as has just been said, of that which is best, so that the form of the future creation may itself be best: it is evident that by those laws which God Himself in his goodness prescribes for Himself, the only thing of which He could adopt the Idea for establishing the universe is His own essence.[7]

Although Kepler later came to revise the polyhedral doctrine presented in *The Secret of the Universe*, he never wavered in his conviction that mathematical proportions, when properly defined, reveal the secret of the world. It remained axiomatic for the early modern thinker that the divine being designed the world in accordance with the natures of quantities, which, during the creation, he disposed in the most regular, symmetrical, and perfect of relations. This was a principle that Kepler found clearly stated in the commentary on the *Elements* by the Pythagorean and Platonist mathematician-philosopher of late antiquity, Proclus, which Kepler cited with considerable approval: "Mathematics contributes things of the greatest importance to the study of nature, both revealing the ordering nature of the reasoning, in accordance with which the whole has been constructed, and so on, and showing that the simple and primary elements, by means of which the whole of the heaven was completed, having taken on the appropriate forms among its parts, are connected together with symmetry and regularity."[8]

Yet Kepler could also find in Copernicus a basis for the belief in the perfect proportions of the cosmos. Introducing his great treatise, *De revolutionibus*, Copernicus assured Pope Paul III that its findings resulted from a pious veneration of the best of all artifacts. "For a long time," Copernicus wrote in his dedicatory letter, "I reflected on the confusion in the astronomical tradition concerning the derivation of the motions of the universe's spheres. I began to be annoyed that the movements of the world machine, created for our sake by the best and most systematic artisan of all, were not understood with greater certainty by the philosophers, who otherwise examined so precisely the most insignificant trifles of this world."[9] The procedure of the old natural philosophers, Copernicus remarked, not without polemic, resembled that of someone who "collected hands, feet, a head, and other members from various places, all fine in themselves, but not proportionate to one body, and no single one corresponding in its turn to the other, so that a monster rather than a man would be formed from them."[10] Such a practice, he concluded, could not but disfigure the "main point" of true astronomy: "the shape of the world and the fixed symmetry of its parts."[11]

For Kepler, self-declared follower of Copernicus and self-styled reincarnation of Pythagoras, that cosmic "shape" could not but be mathematical, and the "symmetry" of its parts, therefore, could not but be quantitative in form. Yet Kepler insisted on one principle that the disciples of the master of Samos would hardly have granted: according to the early modern metaphysician, the fundamental elements of the world are geometrical, not arithmetical, in nature. Like his Italian contemporary, Galileo Galilei, albeit for his own reasons, Kepler held the universe to be a book written in a "mathematical language" whose "characters" are "different from those of our alphabet" on account of consisting of "triangles, squares, circles, spheres cones and other mathematical figures."[12]

Kepler advanced this argument in theological as well as scientific terms. At the opening of *The Secret of the Universe*, he explained that the act of creation related by the Bible was, in essence, an act of revelation: in fashioning the world, God imprinted an image of himself in all created things. That image, Kepler affirmed, is quantitative in nature. Kepler recalled that not only Aristotle, but also Copernicus held the world to be spherical in form. Is the sphere not the most perfect of figures? Invoking Nicolas of Cusa's distinction between the "curved" and the "straight," Kepler observed that the sphere, in its structure, conjoins these two basic quantities. Moreover, he noted, the sphere can be taken as the illustration of the doctrine of the Trinity, "with the Father at the center, the Son at the surface, and the Holy Ghost in the equality of the relation of the center to the circumference."[13] The study of numbers could not hope to grasp the divine nature of such a figure.

In a lengthy letter to Mästlin from 1595, Kepler explained the matter fully. "Before the creation of the world," he wrote, "there was no number, except [that of] the Trinity, which is God himself. Now, if it is said that the world is created according to the measure of numbers, then this is so in the sense in which it is created according to the measure of quantities. But there are no numbers in the line, nor are there any on a surface." "Number," he concluded, "is an accident [or property] of [geometrical] quantity."[14] Multitudes have no significance if not as the measurements of real, continuous, and created bodies. If arithmetic has a foothold in the creation of the world, it is because in certain points and relations, corporeal beings may be susceptible to numerical representation. Yet the underlying reality of the world remains geometrical and, therefore, only in part definable in the discrete units of arithmetic.

Kepler was well aware that his theory of quantity was hardly traditional. The classical doctrine of mathematical entities had rested on the strict opposition between two fundamentally distinct types of quantity: the continuous and the discontinuous, magnitude and multitude, figures, in short, and numbers. It was this understanding that the early modern thinkers of mathematics revised. Like Simon Stevin and John Wallis, albeit in a form of his own, Kepler maintained that there must be, in fact, but a single *quantum*: the continuous, the old object of geometry. Arithmetic, of course, remained a discipline of great utility, for Kepler no less than for his contemporaries, and he himself brought the art of calculation by number to new heights, most of all in the *Nova stereometria*, where he made an unprecedented use of infinitesimals to define the dimensions of wine barrels.[15] But his theory of mathematics expressly excluded the possibility that arithmetic might possess an ideal object of its own, distinct in nature from that of geometry. The art of numbers was, for Kepler, no more than an instrument for the notation of continuous quanta. When judged with respect to the magnitudes of geometry, "numbers," he wrote, with some severity, "are at a second remove, in a sense, or even at third, and fourth, and beyond any limit I can state, for they have in them nothing which they have not got either from quantities, or from other true and real entities, or even various products of mind."[16] "Arithmetic," he maintained, is "nothing [...] but the expressible part of geometry."[17]

A learned reader of the works of the Pythagorean tradition, Kepler knew well that one domain of nature had long been taken as proof of the presence of numbers in the created world. That domain, of course, was *musica*, or harmonics. Authors from Boethius to Zarlino had maintained that consonances are pleasing because of their arithmetical essence, because in them, and perhaps in them alone, ideal multitudes become perceptible in sensuous and earthly form. From his earliest to his last works, Kepler dissented, and nowhere more forcefully and systematically than in his final work of scientific and philosophical innovation, *Harmonices mundi libri V*, or *The Harmony of the World*, published in Linz in 1619. Here Kepler recalled the classical account of the origin of the theory of musical consonance:

> It is said indeed that Pythagoras was the first, when he was passing through a smithy, and had noticed that the sounds of the hammers were in harmony, to realize that the difference in the sounds depended on the size of the hammers, in such a way that the big ones gave out low sounds, and the little ones high sounds. Now as a proportion is properly speaking observed between sizes, he measured the hammers, and read-ily perceived the proportions at which consonant or dissonant intervals occurred, and melodic or unmelodic intervals occurred between notes. Indeed he passed at once from the hammers to the length of strings, where the ear indicates more exactly what fractions of the string are consonant with the whole, and which are dissonant with it.[18]

Kepler granted that Pythagoras had invented the principle of harmony in recognizing that an acoustical phenomenon could be derived from a mathematical proportion, that "differences in sounds," in short, "depend on size." Yet Kepler would not accept that the sizes in question were numerical in form. He argued that an excessive dependence on arithmetic had led the disciples astray. "The Pythagoreans were so much given to . . . philosophizing through numbers," he observed, "that they did not even stand by the judgment of their ears, though it was by their evidence that they had originally gained entry into philosophy; but they marked out what was melodic and what was unmelodic, what was consonant and what was dissonant, from their numbers alone, doing violence to the natural prompting of hearing."[19] Erroneously holding the Platonic "remainder" (256:243) to be melodic, confusedly believing the interval of the "minor tone" (10:9) to be unmelodic, the followers of Pythagoras thus came to betray the principles that their venerated master had bequeathed to them. Ptolemy, we learn, was the first to grant the ears their rightful place in music, emending the Pythagorean system in accordance with the realities of hearing.[20] Yet he, too, succumbed, albeit in his own way, to "abstract numbers," distorting the true nature of intervals in his *Harmonics*. It was on account of his faith in arithmetically measured multitudes, Kepler reasons, that Ptolemy "denied that the thirds and sixths, minor and major (which are covered by the proportions 5:4, 6:5, 5:3 and 8:5) are consonances, which musicians of today who have good ears say they are."[21]

Had the followers of Pythagoras been faithful to their old master, Kepler argued, they would, instead, have trusted in their original "judgment of the ears." They would have observed that like all other physical estimations, this judgment bears on bodies and, therefore, on quantities subject to the rules of geometry. Kepler argued that the study of figures, in fact, furnishes the sole basis for the resolution of the problem that had troubled the theory of harmony since antiquity: the problem, namely, of the distinction between consonance and dissonance. Ancient harmonic theory since Euclid, if not before, had held that certain principles of numbers explained the difference between pleasing and displeasing sounds.[22] In his *Harmonic Institutions* of 1558, Zarlino, despite his innovations, had upheld that old position. Although he had argued that the basic "harmonic numbers" were not four, as the ancients had believed, but rather six, the Italian theorist had continued to maintain, like Boethius before him, that musical intervals express the properties of a limited set of measured multitudes. On this matter, Kepler found Zarlino's doctrine of *musica* hardly more convincing than the ancient. Kepler maintained that both failed to understand that numbers, being "but the expressible part of geometry," could not explain the nature of the created world. Had not the ancients themselves, in their measurements, availed themselves of continuously extended bodies? The monochord, Kepler recalled, defines intervallic relations by reference to the lengths of strings. Numbers come later, when theorists express sonorous extensions in terms of measured multitudes. "Since the terms of the consonant intervals are continuous quantities," Kepler explained, "the causes which set them apart from the discords must also be sought among the family of continuous quantities, not among abstract numbers, that is, in discrete quantity."[23] Geometry alone can offer reasons for the facts of music.

In book 3 of *The Harmony of the World*, Kepler set out to accomplish, for this reason, what centuries of hapless Pythagoreans had failed to do: to present a complete theory of music based on not arithmetic but geometry. He had attempted to offer such an account as early as 1596, when he argued in *The Secret of the Universe* that the laws of consonance and dissonance might be derivable from the relations between the five regular Platonic solids.[24] In 1619, Kepler conceded that that account had failed, but he added that his basic intuition about the correlation between music and geometry had been correct. "Although I remarked at a fairly early stage that the causes [of harmonies] must be sought in the plane figures, and you see the seeds of the matter already scattered in the Chapter referred to, XII, of *The Secret*," he wrote, "yet they racked me exceedingly for a long time, before all my mind's doubts were satisfied."[25] This occurred once Kepler realized that the geometrical beings capable of revealing the principles of harmony are not three-dimensional but rather two-dimensional figures. Surfaces, not solids, can resolve the ancient question of the "causes of harmony."

The basic doctrine presented in *The Harmony of the World* may be easily stated. Bending a single string so that its beginning and ending meet, Kepler explains, one obtains a circle. One may then inscribe a regular polygon within it. Such a figure will necessarily divide the circumference of the circle into a certain number of equal arcs: an inscribed triangle, for example, will produce three arcs; an inscribed square will produce four; a pentagon, five; and a hexagon, six. Kepler now suggests that one distinguish between two varieties of such arcs. "Parts," he argues, may be told apart from "residues." Wherever one or more of the sides of the polygon subtends one or more arcs, such that the sum of the arcs does not surpass half of the circumference, Kepler speaks of a "part" or "parts." The remaining portion of the circle, then, will be called a "residue." Between part and whole, between residue and whole, and between part and residue, three new relations may then be established.

Through an exposition in terms both musical and geometrical, Kepler advances a major thesis: the relations between the wholes, parts, and residues of such polygons inscribed in circles define all the musical consonances admitted by "musicians of today who have good ears."[26] Moreover, such relations define consonances alone, in distinction to dissonances. In short, polygons inscribed in circles succeed, for Kepler, where numbers fail. Figures reveal the true and the only true mathematical basis of musical sounds. Yet in Kepler's method of geometrical representation, certain rules are required. As H. F. Cohen has shown, they can be reduced to three axioms.[27] First, the only admissible polygons are those that may be directly constructed by compass and ruler.[28] Second, "replicas are eliminated because of the apparent identity of two notes one octave or more apart from each other."[29] Third, "the section of a circle by a regular polygon generates consonant proportion if and only if none of the proportions part to whole, residue to whole or part to residue could before have been reached through an unconstructable polygon, like, for instance, the heptagon."[30]

The last axiom is in some sense the least evident, yet the most important. It points to one mathematical fact to which Kepler accorded great significance: were certain polygons, such as the heptagon, inscribed within the circle, they would display lengths that remain strictly incommensurable with the diameter. Such figures, to use Kepler's technical term, are "indemonstrable," that is, unconstructable, for they cannot be defined in the geometrical sense of the word. The opening propositions of *The Harmony of the World*, which owe much to Euclid, are explicit on this point. Mathematical knowledge is limited by nature: "In geometrical matters," we read, "to know is to measure by a known measure, which known measure in our present concern, the inscription of figures in a circle, is the diameter of the circle."[31] If the side of a polygon cannot be defined by ruler and compass with reference to the diameter, then it cannot be "known." Kepler comments that it would be imprecise to name such a magnitude "irrational," "inexpressible," or "surd," since "there are many lines, which, although they are Inexpressible, are defined by the best computations," just as some quantities often considered "surds" can be expressed in square proportion. Yet an inscribed line that cannot be exhausted by any number of aliquot parts of the diameter remains absolutely unknowable. Posing a barrier to geometrical construction, it furnishes the sole mathematical criterion for the distinction between consonance and dissonance: all ratios exhibited in the arcs cut off by such polygons can and, indeed, must be excluded from the domain of harmony.

It would be an error to consider that domain, for Kepler, to be confined to man. More than once, Kepler stated that geometrical necessities are also divine. From a theological perspective, they are, as he repeatedly affirmed, "coeternal." "Geometry is coeternal with God," he maintained in *The Harmony of the World*.[32] As Jean-Luc Marion observes, here, "*coeternal* means more than merely *eternal*."[33] It suggests grounds for substantial comparison. Mathematical laws and theorems are for Kepler not only older than the time of all created things; they are also similar to God on account of the "eternity" they share with him.[34] Laying bare the principles of harmony, Kepler thus presents the original Ideas that presided over the creation: the "archetypes" from which the universe was made. Their rules and their limitations are common to the human mind and the divine: what is "indemonstrable" to man remains equally "unknowable," then, to God.

Yet there is more. To understand the truths of mathematics, for Kepler, is not only to grasp the original conceptions of the Artisan. It is to grasp them as he does so himself and also to do so according to his will, since "God wished that man, who is His image, might have an understanding in common [*intellectum... secum comunem*] with Him."[35] Decidedly, if implicitly, the old theory of an analogy between the human and the divine mind gives way before a new doctrine of univocity: in the act of grasping the geometrical truths of harmony, the created and the uncreated intellects come to "know" the same objects in the same sense. As Gérard Simon writes, mathematics thus shows itself to be "the mediator between man and God [....] It is the only language they both possess. The truth of demonstrations holds even for the creative intelligence, and the beauty of proportions that lends the world its harmony is as charming to the astronomer who calculates it as it is to the supreme being who contemplates it."[36]

Only if one grasps this conception of mathematics can one measure the importance that the laws of harmony acquire in Kepler's geometrical cosmology. The early modern thinker was certainly willing to argue that the human mind finds certain intervals pleasing on account of accidents of sensible beings. "The actual sensible part of harmony, as harmony," he conceded, "is an accident of sensible things, just as, of course, it is an accident of the same things to be seen and heard and so forth."[37] But he believed sensible harmonies could exert no effect upon us had they not the form of "abstract emanations of sensible things," which, while "entering through the senses," reach the "tribunal of the soul," which judges them to be proportionate or disproportionate according to its rules.[38] Yet audible consonances, in any case, fill only part of *The Harmony of the World*. They reflect but a fraction of the mathematical necessities inscribed in creation. Beyond the realm of sound, beyond the domain of sensation, and even beyond this Earth, "pure and secret" proportions, the physicist taught, may be discerned. This much follows from the principle that it is not right that "the best should make anything except the most beautiful," in accordance with the most perfect of coeternal ideas: all nature, for Kepler, reveals ideal and intelligible proportions. "For the Creator," he explained, "the actual fount of geometry, who, as Plato wrote, practices eternal geometry, does not stray from his own archetype."[39]

At first glance, one might consider such an idea to represent the remnant of an ancient belief, as Kepler's own citation of Plutarch citing Plato suggests.[40] Yet it was precisely this affirmation of the universality of geometry that cleared the way for Kepler's contribution to the emergence of early modern cosmology, at least in the two fundamental aspects that Alexandre Koyré long ago attributed to it. Precisely in defining the world as the artifact of a "geometricizing god" whose harmonies are everywhere equally ideal, regular, and mathematical, Kepler accomplished two basic steps of the "scientific and philosophical revolution": "the destruction of the Cosmos, that is, the disappearance, from philosophically and scientifically valid concepts, of the conception of the world as a finite, closed, and hierarchically ordered whole," and its replacement by the idea of a physical universe "bound together by the identity of its fundamental components and laws, and in which all these components are placed on the same level of being."[41] Already in *The Secret of the Universe*, Kepler had in fact dismissed as "monstrous and absurd" the old idea—which Copernicus may have himself espoused—that beyond the moon, real spheres move through the sky, carrying the stars.[42] Admittedly, Kepler never doubted that at the farthest point of remove from the Earth lay "fixed stars" that enclose the spherical universe within them. But unlike his predecessors, Kepler refused to grant that in the created universe, differences in physical position imply any fundamental differences of nature. "It is Kepler,"

Koyré wrote, "who frankly and consciously inaugurates a modern conception of the essential identity of the elements that compose the world"; it is he who unifies "terrestrial physics" and "celestial physics" in the invention of a single science of nature, recognizing only "a perfectly homogenous space, which is entirely geometricized, and in which 'places' are rigorously equivalent."[43] This "inauguration" followed from the postulate of a god who, from the heavenly bodies down to the single snowflake, could not stray "from his own archetype." For Kepler, "a perfectly homogenous space" was, in short, the correlate of a creator to whom mathematics were absolutely "coeternal."

It is hardly surprising, therefore, that the fifth and crowning book of *The Harmony of the World* should take leave of all audible and sensible phenomena, concentrating instead on mathematical proportions legible in the planets alone: "The Most Perfect Harmony of the Heavenly Motions," as the title has it, "and the Origin from the Same of the Eccentricities, Semidiameters, and Periodic Times."[44] Here Kepler offers what may be his fullest account of the regularities of the celestial bodies, defining the laws of planetary motion that still bear his name. These are generally considered to be three in number. Kepler presents the First Law, known as the law of the elliptical orbit, in two propositions, which allude to the findings he made in his *Commentaries on Mars*: "It was shown by me that *the orbit of a planet is elliptical, and the Sun*, the fount of motion, *is at one of the focuses of that ellipse*."[45] The Second Law, known as the "distance" or "area law," states that in its orbit, a planet sweeps through equal domains in equal time. Given the elliptical form of the planetary orbit, this principle implies that the closer the planet is to the Sun, the more swiftly it will move. In Kepler's own terms: "A planet's unequal delays in equal parts of the eccentric follow the proportion of its distances from the sun, the source of motion."[46] Finally, the Third Law dictates that for any two planets, the proportion of mean orbital times is exactly one and a half times the proportion of mean diameters.[47] Kepler appears to have discovered this principle relatively late in his work on *The Harmony of the World*, which, while published in 1619, was conceived in the 1590s.[48] Introducing the Third Law, Kepler remarks:

> I thought of it on March 8 of this year 1618, but was unsuccessful in my attempt to verify it numerically, and rejected it as untrue. Then on May 15, it came back to me and with a new assault drove the obscurity out of my mind. Seventeen years' work with Tycho's observations and my present meditations on this subject were in such good agreement that at first I thought I was dreaming and was assuming the truth of what I was going to find. But it is most certain and most exact that the proportion between the periods of any two planets is precisely three halves the proportion of the mean distances, that is, of their orbs.[49]

It is this astronomical principle that is generally termed "the Harmonic Law," and for a simple reason. The relation it establishes between time and distance for two given planets is that of a fundamental arithmetical inequality well known in the study of music: the ratio, namely, of three to two (3:2). In other words, for two planets moving elliptically about the Sun, times and distances exhibit one basic mathematical consonance: the interval of the perfect fifth.

It would be difficult to overstate the importance of the Harmonic Law for Kepler's cosmology, even if, as J. V. Field observes, it does not play "a large part" in the astronomical calculations of book 5 of *The Harmony of the World*.[50] The first two laws establish that strictly geometrical regularities dictate the paths and speeds followed by each individual planet in its orbit around the Sun. Yet the Third Law does much more. It succeeds in relating one planet to another, such that between any two celestial bodies, an exact harmony may be observed. Discovering this law, Kepler believed himself to have uncovered a physical and mathematical principle that commands the form of the universe as a whole. "The discovery for the sake of which I applied the best part of my life to astronomical studies, I visited Tycho Brahe, and I chose Prague as my seat," he now declared,

> that discovery at last, with God the Best and Greatest, who had inspired my thought and had aroused this mighty ambition, also prolonging my

life and the vigor of my talents, and supplying the rest of requirements through the generosity of two Emperors, and the chief men of this province of Upper Austria, on the completion of my previous work in the province of astronomy to a sufficient extent, at last, I say, I have brought that discovery into the light, and have most truly grasped beyond what I could ever have hoped: that the whole nature of harmony, to its full extent, with all its parts...is to be discovered among the celestial motions.[51]

Kepler rose to the heights of inspired exaltation, certain that he had achieved that for which God had waited six thousand years. "It is my pleasure," he concluded,

to yield to the inspired frenzy; it is my pleasure to taunt mortal men with the candid acknowledgement that I am stealing the golden vessels of the Egyptians to build a tabernacle for my God from them, far, far away from the boundaries of Egypt. If you forgive me, I shall rejoice; if you are enraged with me, I shall bear it. See, I cast the die, and I write that book. Whether it is to be read by the people of the present or of the future makes no difference: let it await its reader for a hundred years, if God Himself has stood ready for six thousand years for one to study him.[52]

Yet Kepler developed his musical account of the universe further. Having established, by the Third Law, the physical basis for a consonance between celestial bodies, he proceeded to show that the First and Second Laws, correctly understood, also imply the existence of a harmony among the planets. It suffices to recall the rules of the elliptical orbit and variable planetary speeds. They dictate that each celestial body moves between two extremes of velocity: that attained at the point of closest proximity to the Sun, the perihelion, where planetary speed is greatest, and that reached at the aphelion, where it is least. If one puts these two values in relation, they will naturally form an inequality. More precisely, they will form a ratio, like two sections of a single monochord. Kepler argued, with copious astronomical documentation, that if one considers the extreme orbital speeds of the six planets known to early modern astronomy, one fact can hardly be denied: between aphelion and perihelion, the planets traverse intervals that, when rendered mathematically, are musical in form. Mercury, moving between the extreme velocities of twelve and five (12:5), passes through the relation that, when referred to string lengths, produces an octave plus a minor third. Venus, in the movement between the values of twenty-five and twenty-four (25:24), exhibits the inequality that defines the chromatic half tone. The Earth, between speeds of sixteen and fifteen (16:15), makes a diatonic half tone. Mars, whose variable motions pass between three and two (3:2), gives rise to a perfect fifth. The velocities of Jupiter, passing from six to five (6:5), exhibit a minor third. Saturn, finally, in its transit from perihelion to aphelion, moves from the value of five to that of four (5:4), thus producing a major third.

Kepler's astral harmonies, however, are richer than this summary allows. One may begin by observing that the variation in velocity exhibited by a planet in its orbit can be precisely correlated with an interval known to musical theory. But between the aphelion and perihelion of one planet and those of another, further relations may be established. Harmonies, then, can also be discerned not only in the course of each celestial body, but also in their various combinations. "Further," we read,

> there is a great distinction between the harmonies which have been set out between individual planets, and between planets combined. For the former cannot exist at the same moment of time, whereas the latter can absolutely; because the same planet when it is situated at its aphelion cannot at the same time also be at its perihelion which is opposite, but of two planets one can be at its aphelion and the other at its perihelion at the same moment of time.[53]

A learned reader of modern as well as ancient harmonic theory, Kepler quickly drew the conclusion that such a musical structure implied. While each celestial body sings a one-part melody of the kind familiar to the ancients, the six planets, taken together, produce a "melody" of another sort: "a melody of several voices, called figured, and the invention of recent centuries," that is, more simply, polyphony.[54] If, by the art of musical notation, one then sets the various planetary intervals in the space of a single tonal expanse, assigning one note to the slowest of planetary motions and relating the swifter ones to it, one obtains a musical transcription of all the soundless heavenly movements.[55] In book 5 of *The Harmony of the World*, Kepler thus obtained six "scales," which, in contrapuntal "combination," "match modern figured music."[56] Copernicus had likened the movements of the planets to an elaborate ballet.[57] Kepler now suggests another model, based on observation and exact calculation. Once heaven has been cast as the page of a well-staffed polyphonic score, the planets may be perceived as executing a cosmological composition in six parts. "Therefore, the motions of the planets," he concluded, "are nothing but a kind of perennial harmony (in thought, not sound), through dissonant tunings, like certain syncopations and cadences (by which men imitate those natural dissonances), and tending towards definite and prescribed resolutions, individual to the six terms (as with vocal parts) and marking and distinguishing by those notes the immensity of time."[58]

Yet Kepler's strictly geometrical representation of the universe also entails consequences that are less easily reconcilable with the ideal of harmony. It is worth noting, first of all, that in Kepler's account, the astronomer may grasp the proportions of the world only if he abstracts himself from his own position, imagining that he may look—or hear—from a perspective that he cannot occupy. It is certain that the harmonious disposition of the planets cannot be grasped from the Earth, for they are intelligible solely from the vantage point of the Sun. To this degree, Kepler's account of the order of the universe recalls that of Copernicus, according to whom, as Fernand Hallyn observes, "God has created a 'symmetrical' universe, but this 'symmetry' only appears to man if he assumes that he is not viewing things from the center."[59] Kepler was well aware of this fact: to perceive the musical ratios that obtain between the planets, one must perform a curious experiment. First, "one must," as he noted, almost in passing, "place an eye on the sun."[60] Second, to hear the harmonies in their succession and simultaneity, one must abstain from "applying to the celestial concert our scale of time," for the music of the heavens unfolds uninterruptedly from the act of creation to the Last Judgment.[61]

Yet even from a solar and eternal perspective, oddities in the order of the world are audible. The most striking among them derives from the fact that although the minimum and maximum planetary speeds for each planet are discrete, like the two terms of a musical interval, in its orbit, a planet must traverse an infinitely divisible continuum of velocities, as Kepler himself remarks.[62] "Each planet has its own scale, determined by its extreme speeds," D. P. Walker writes, yet its transition "from the lowest to the highest note and back again is not really articulated into steps of tone and semitone, as in a musical scale, but represents a continuous acceleration and deceleration of the planet's speed, so that, if they actually emitted sounds (which they do not), their 'scales' would sound like a siren giving an air-raid warning."[63] "Musically, this means," Michael Dickreiter observes, "that the interval that is defined by the two extreme speeds is realized a glissando."[64] By virtue of its traditional form, Kepler's musical notation remains in this sense more arithmetical than the quantitative reality it grasps. The truth is that between aphelion and perihelion, there lie magnitudes, not multitudes, of quantity: continuous modulations to which, in acoustical terms, incessantly shifting spectra of sound alone would correspond.

A greater disharmony, however, threatens Kepler's polyphonic universe. His cosmology can hardly avoid confronting one basic, yet troubling question, which bears on the place in which figures, solids, lines, and curves, are drawn. Such a query concerns less the archetypes themselves than the area of their inscription, that in which, in other words, the Platonic God always geometricizes. It cannot be doubted that Kepler's doctrine of harmony requires that the Sun be at its center of the universe. But can one be certain that the universe has a center? What if it were extended like an endless line, or an area without limits? Such a possibility, however perplexing it might seem, was by the seventeenth century all too easily imagined.

Ancient philosophers had debated whether there could be such a thing as a limitless place or body, and whether it might be identified with this world. Aristotle considered the matter in detail in Book 1 of his treatise on the heavens. Since he maintained space to be no more than the surrounding of bodies, he reduced the question of an infinite space to that of an infinite body.[65] This, in turn, he believed he could answer definitively in the negative. He began by noting that all bodies are made of either simple or composite constituents. Bodies made of a limited number of simple constituents must be limited, "since what is composed of things limited both in number and in size is itself limited in number and size; for it is as great as the things of which it is composed."[66] Aristotle implicitly excluded the possibility that any body might be composed of simple elements that are, in themselves, unlimited. Yet it still remained for him to decide whether the universe might itself not be a simple and unlimited body. Since he assumed that the nature of the universe could be inferred from the heavens, and since he took as self-evident that the heavens turn in a circle about the Earth, he posed the question of an infinite world in the following form: Can an unlimited body move in a circular fashion? Aristotle then advanced six physical and astronomical arguments that excluded such an eventuality. For example, if an unlimited body were to turn, it could do so only in a circle that was itself unlimited, but "the unlimited cannot be traversed," and so the unlimited also cannot be traversed in a circle.[67] One after another, the Aristotelian demonstrations led to a single conclusion: a body in circular revolution must be "limited as a whole" (πεπεράνθαι πᾶν).[68] "It is evident," Aristotle concluded, "that a body that moves in a circle is neither endless [ἀτελεύτητον] nor unlimited [ἄπειρον] but must, rather, have some end."[69] This thesis was also in accord with Aristotle's conception of the world as composed of a series of concentric circles that must reach a limit.

Aristotle's proofs held sway for centuries. Their validity, however, became uncertain when Ptolemaic cosmology fell into doubt. As Jean Seidengart notes, "from the moment Copernicus definitively abandoned geocentric and geostatic appearances, this entire means of assessment collapsed on its own."[70] One possibility, then, could no longer be avoided: one might well conceive the universe to be indefinitely and even infinitely extended, like the space of Euclidean geometry. It was increasingly recalled that even after Aristotle, in fact, certain thinkers had held the world to be unlimited in magnitude. Some, contesting the link Aristotle had established between place and bodies, argued that the universe might well include a boundless empty space: that is, a void. Others, instead, dismissed Aristotle's refutation of the idea of an unlimited heaven. The atomists of antiquity were doubtless the most famous advocates of the idea of a limitless universe, but Leucippus, Democritus, Epicurus, and Lucretius were not the only philosophers who entertained the possibility that our particular cosmos might be exceeded by a "whole" or "universe" (τὸ πᾶν) that had no boundaries.[71] It has been observed that "Xenophanes considers the earth as 'rooted in the infinite space,' Seleucus affirms that the world knows no limits, and both Diogenes and Archelaus regard the universe as infinite."[72] And in his book on the face of the moon, Plutarch did not hesitate to declare: "The universe is infinite; and what is infinite has neither beginning nor limit, so that it does not possess a middle; for infinity is the deprivation of limits."[73]

Similar affirmations could be found among the late Scholastic thinkers, who advanced their theses with rigor and audacity. Nicole Oresme, in his *Quaestiones* on Aristotle's book on the heavens, argued for the possibility of a physical void beyond the spheres: "outside the heavens there may be a vacuum," he remarked, "since God can create a place or a body there."[74] In his commentary on the *De caelo* in Middle French, he went further, attributing such immensity to a kind of space:

> Outside the heavens, then, is an empty incorporeal space [*une espace wide incorporelle*], quite different from a plenum or corporeal space, just as the extent of this time called "eternity" is of a different sort than temporal duration, even if the latter were perpetual.... Now this space of which we are speaking is infinite and indivisible [*infinie*

et indivisible], and it is the immensity of God and God Himself, just as the duration of God called eternity is infinite, indivisible, and God Himself.[75]

Perhaps under the influence of Oresme, the Jewish Catalan philosopher of the fourteenth century Hasdai Crescas developed a physics that broke even more sharply with Aristotle. For the author of the *Light of the Lord*, "the universe is not that finite system of concentric spheres of Aristotle's conception but rather the infinite vacuum within which Aristotle's finite universe is contained as in a receptacle."[76]

Copernicus himself had encountered this possibility for reasons that followed from his astronomy. Since antiquity, it had been known that heliocentric astronomy necessarily entails a consequence: when the Earth is no longer imagined to lie at its center, the universe swells greatly in its dimensions. Because of the extreme difficulty of observing a stellar parallax in a heliocentric cosmos, the dimensions of the universe grow strictly immeasurable to the eye. Citing Aristarchus of Samos, Archimedes showed, however, that this much can be established: if the planets move about the Sun, then the distance of the Sun to the fixed stars is incalculably greater than the distance of the Earth to the Sun.[77] Copernicus knew this well. Nonetheless, he chose not to pronounce himself on the question of the exact dimensions of the cosmos. "This," he noted prudently, alluding to the impossibility of establishing a parallax in a heliocentric universe, "establishes no conclusion other than the indefinite magnitude [*indefinitam magnitudinem*] of the heavens as compared to the earth. As to how far this immensity [*immensitas*] stretches, this is not at all clear."[78] The astronomer refrained from drawing any more cosmological conclusions: "Let us leave it to the disputations of the philosophers [*physiologi*] to decide if the world is finite or infinite; we are in any case certain that the earth, between its poles, is limited by a spherical surface."[79]

Those who dared to resolve the question of the size of the universe in favor of immensity and infinity were the younger exponents of Copernicanism: Francesco Patrizi, Giovanni Battista Benedetti, Rainiarus Ursus, John Dee, Thomas Digges, William Gilbert, and Giordano Bruno.[80] None among them was more eloquent than Bruno, whose works, invoking Democritus and Epicurus against Aristotle, proclaimed the finite world long defunct. "Even many of the Aristotelians," Bruno wrote, "could not accept Aristotle's argument against the void."[81] Properly understood, Bruno argued, space is a continuous physical magnitude that precedes all bodies and impassively receives all things, being unmixed and impenetrable, formless, placeless, immobile, uncontainable, and incomprehensible, neither substance nor accident.[82] Space is what extends itself through a universe possessing the characteristics of being infinite in its magnitude and innumerable in the multitude of its worlds. "We say," Bruno wrote, "that there exists an infinity, and it is an immense ethereal region in which there are innumerable and infinite bodies like the Earth, the Moon and the Sun, which we call worlds, composed of plenum and void."[83]

Kepler was well aware that certain followers of Copernicus had claimed the universe to be infinite. He was himself an interested, if critical, reader of William Gilbert and Giordano Bruno. Moreover, among his personal acquaintances and correspondents were open advocates of the idea of an unlimited cosmos, such as Johann Matthäus Wackher von Wackhenfels and Edmund Bruce, who, in a letter to Kepler from 1603, declared: "I believe there are infinite worlds [*mundos esse infinitos*], each of them, however, being finite, just as [that] whose middle point of the planets is the centre of the Sun."[84] Kepler chose to address the question publicly after 1604. That was the year of an important supernova, which suddenly brought a "new star" to the attention of stargazers across Europe. Kepler noted that some had chosen to infer from this event the most far-reaching of conclusions. In a work on the question published two years later, he wrote:

> There is a sect of philosophers who...do not start their ratiocinations with sense-perception or accommodate the causes of the things to experience. Instead, they immediately and as if inspired by some kind

of enthusiasm conceive and develop in their heads a certain opinion about the constitution of the world; once they have embraced it, they stick to it; and they drag in by the hair [things] which occur and are experienced every day in order to accommodate them to their axioms. These people want this new star and all others of its kind to descend little by little from the depths of nature, which, they assert, extend to an infinite altitude, until according to the laws of optics it becomes very large and attracts the eyes of men; then it goes back to an infinite altitude and every day [becomes] so much smaller as it moves higher.[85]

Kepler again considered the idea of an infinite universe in his last major work, the *Epitome of Copernican Astronomy*, whose volumes appeared between 1615 and 1621. Each time, his position was clear: he judged the hypothesis inadmissible. Kepler granted that after Copernicus, the classical materialist conception of an unlimited cosmos could again be proposed, despite Aristotle's refutation of it. Alluding to the ancient philosopher's demonstrations of the impossibility of conceiving an infinite body in rotation around the Earth, Kepler observed that "since Copernicus has suppressed that movement, one can entertain the idea that the sphere of fixed stars is infinite," as did "a school of pagan philosophers in antiquity."[86] Yet such theory, he maintained, could also once again be dismissed.

Kepler invoked several arguments against the idea of an infinite universe. He drew a first set of objections from the philosophical doctrine of infinity developed by Aristotle and his Scholastic commentators. Aristotle had taught that everything that can be said to "be" is either in actuality or in potentiality.[87] Infinite Being, he maintained, can be only of the second variety. One may, for instance, speak of infinite division or infinite addition, yet only with reference to procedures that have yet to be accomplished, for in fact, every quantity is finite. In an early philosophical book written for the purpose of teaching mathematics, *On Quantities*, Kepler clearly propounded this traditional account: "As to the actual infinite," he wrote, "we deny with Aristotle *Phys.* III, 5 that there is any actual infinite quantity, either sensible or intelligible."[88] Of course, Kepler conceded that "it is always possible to think of something not only greater than any number, but greater than any magnitude, to infinity." But the idea of such infinity can be realized only by the thought of successive addition or subtraction, not in a single concept, and such an idea can in any case never be tied to existing magnitudes, for "nothing in nature is actually infinite."[89] Kepler's later statements on the dimensions of the cosmos perfectly respected this reasoning. "Whoever declares the sphere of fixed stars infinite commits a *contradictio in adjecto*," he wrote in *De stella nova*. "In truth, an infinite body cannot be conceived by thought, since the concepts of the mind regarding infinity refer either to the meaning of the word 'infinity' or to something that exceeds all conceivable numerical, visual, or tactile measure: that is to say, something that does not exist in actuality. For an infinite measure cannot be conceived."[90]

Kepler also raised a second set of objections to the idea of a boundless universe. These involved the methods and findings of the science of astronomy. He pointed out that the study of the heavens advances by observation. It is by scrutiny of the sky that astronomers come to discern the unifying principles that underlie apparent diversities, such that "phenomena," as the old phrase has it, may be "saved" in becoming intelligible. Yet how might one observe a cosmic infinity, and how might it explain anything? The belief in a boundless universe could lead only to epistemological absurdities. Kepler remarked that if one holds the universe to be limitless, one will necessarily claim that the fixed stars, while bounded in their relation to the Earth, are otherwise unbounded. "Yet can it be believed," he asked, "that, having a limit on one side, they extend without limits on the other? And how could there be a center in infinity, given that in infinity it is everywhere? Any point in infinity is equally, that is to say infinitely, separated from the extremities, which are infinitely distant. It would follow that the same point is and is not a center and many other contradictory things—that one will correctly avoid," he concluded, "if one posits an outer limit to the sphere of the fixed stars, as there is an inner one."[91]

Finally, Kepler argued that the appearance of the heavens

manifestly disproves the hypothesis of a boundless universe. In his *Epitome of Copernican Astronomy*, if not in his *Harmony*, he implicitly called upon a basic rule of modern physics, which would one day be known as the "cosmological principle": the universe must present the same semblance, no matter the position and perspective of the one who observes it. The cosmos, in other words, must be isotropic.[92] Given the principle of the uniformity of nature, Kepler notes, one may wager that were the heavens truly limitless, the placement of stars across them would be manifestly homogenous. Stars of equal numbers, set in equal groups, would follow each other to infinity. The Earth's Sun, then, would fall after one set of such stars and before another. "Among the innumerable places in that infinite assembly of the fixed stars, our world, with its sun, would be one place in no way different from other places around other fixed stars, as represented," Kepler wrote, pointing to a curious illustration, "by the adjoined figure *M*."[93] But this, of course, is not the case. Everyone knows that the stars fill the sky in the most varied of constellations.

For the author of *The Secret of the Universe* and *The Harmony of the World*, there was, however, a further reason to set the idea of a limitless universe aside. It may be that this reason alone ultimately explains the others. It can be simply stated: were the cosmos to be indefinite and even infinite in its dimensions, the possibility of a geometrical cosmology would be excluded. The harmony of shapes and solids cannot admit the unlimited. "From this complete, thoroughly ordered and most splendid universe," Kepler declared at the opening of *The Secret of the Universe*, "let us reject straight lines and surfaces, as they are infinite, and consequently do not admit of order."[94] This was one axiom Kepler stated on the first page of *The Harmony of the World*:

> Shape and proportion are properties of quantities, shape of individual quantities and proportion of quantities in combination. Shape is demarcated by limits, for it is by points that a straight line, by lines that a plane surface, by surfaces that a solid is bounded, circumscribed, and

Kepler's Infinite Universe (*Epitome astronomiae Copernicanae*, 1618–21 (Johannes Kepler, *Gesammelte Werke*, ed. Walter von Dyck and Max Caspar [Munich, 1937–], vol. 7, p. 42).

shaped. This is why that which is limited, circumscribed and figured can also be grasped by the mind. What is infinite and unbounded, as such, cannot be contained by the bounds of any knowledge based on either definition or demonstration.[95]

A limitless universe, in short, would be "indemonstrable," like those polygons that neither the human nor the divine intellect can construct. Boundless, it would be neither geocentric nor heliocentric; it would be acentric. In such a universe, one might, of course, still grant Kepler's astronomical laws. Some planets would move elliptically around some star, sweeping through equal areas in equal time, their mean orbits in stable relation to their mean diameters. But what sense would their aphelia and perihelia, their accelerations and their decelerations, retain in a cosmos of innumerable stars and planetary systems? What music would they make? Evoking the unlimited cosmos in *De stella nova*, Kepler commented: "The mere thought of it brings with it I know not what of secret, hidden horror; one finds oneself wandering in this immensity, which knows no boundaries, no center, and, therefore, no defined places at all" (*Quae sola cogitatio, nescio quid horroris occulti prae se fert; dum errare sese quis deprehendit in hoc immenso; cujus termini, cujus medium, ideoque et certa loca, negantur*).[96]

One might argue that Kepler retreated from the reality he had glimpsed. Having gone perhaps further than any early modern thinker in envisaging the physical structure of the modern universe, he recoiled before the possibility that unlike the cosmos of old, the new world might possess no limits and no center. Perhaps it was on account of his allegiance to Copernicus that he refused to abandon the fundamental point of astronomical orientation that was the Sun. Had the author of the new world system himself not likened the heavens more than once to a "visible God"?[97] "At rest, in the middle of everything," Copernicus had famously written, "is the Sun. For in this most beautiful temple, who would place this lamp in another or better position than that from which it can light up the whole thing at the same time?" Justly, he remarked, it has been named the "lantern," the "mind," the "ruler" of the universe. "Thus indeed, as though seated on a royal throne, the Sun governs the family of planets revolving around it."[98] Modern science, it is often said, remained in its inception partly also theology. Kepler's would then have been a failure necessitated by historical circumstance. One would add that others after him succeeded where he did not, developing an astronomy that could, and did, concede a cosmos shorn of pleasing proportions.

Yet Kepler's "secret, hidden horror" before the infinite also concealed an insight. Kepler discerned one truth that others failed to observe, even as he furtively averted his gaze from it. He saw that the limitless universe is out of this world. This is why, for us, it remains inconceivable, as long as one wishes to define the conceivable with respect to our experience. One might also choose to refer such questions to faculties other than our own. Already in Kepler's life, astronomy had begun to free itself of the bounds of the human body, as Galileo's telescope disclosed findings that far outstripped what the naked eye could see. From that moment onward, one might well argue, the study of the heavens would observe a sky inaccessible to human perception. Increasingly exact instruments of measurement would come to detect new regularities, revealing the forms of hitherto unimagined quantitative relations. One can wager that such discoveries, however, might not affect Kepler's view. His intuition may have been sound. In a universe without limits, its center everywhere and nowhere, its boundless stars distributed in endless uniformity, one might well continue to grasp natural phenomena by mathematical means. But a harmony of the world would not be heard. One might wait another six thousand years, yet no thinker, sage, or scientist, would step again into the forge, and no Pythagoras would be reborn.

Notes

CHAPTER ONE: INTO THE FORGE

1. Boethius, *De musica*, book 1, chapter 10, in Anicius Manlius Severinus Boethius, *Fundamentals of Music*, trans. Calvin M. Bower, ed. Claude V. Palisca (New Haven: Yale University Press, 1989), p. 17; for the Latin text, see Boethius, *Traité de Musique*, ed. and trans. Christian Meyer (Turnhout: Brepols, 2004), pp. 46–50.

2. Boethius, *De musica*, book 1, chapter 10, in *Fundamentals of Music*, p. 18.

3. *Ibid.*

4. *Ibid.*

5. *Ibid.*

6. Aristotle, *Metaphysics* 987b.28.

7. Aristotle, *Metaphysics* 987b.11.

8. Aristotle, *Metaphysics* 986a.18.

9. On Aristotle's reconstruction of Pythagoreanism, see Walter Burkert, *Lore and Science in Ancient Pythagoreanism*, trans. Edwin L. Minar (Cambridge, MA: Harvard University Press 1972), pp. 28–52; cf. Leonid Ja. Zhmud', "'All is Number'?: 'Basic Doctrine' of Pythagoreanism Reconsidered," *Phronesis* 34.3 (1989), pp. 270–92; Carl Huffman, "The Role of Number in Philolaus' Philosophy," *Phronesis* 33.1 (1988), pp. 1–30.

10. On the Pythagorean "fourfold" or "four-group," see Burkert, *Lore and Science in Ancient Pythagoreanism*, pp. 71–73; Armand Delatte, "La tétractys pythagoricienne," in *Études sur la littérature pythagoricienne* (Paris: Champion, 1915), pp. 249–68; Paul Kucharski, *Étude sur la doctrine pythagoricienne de la tétrade* (Paris: Les Belles Lettres, 1952).

11. Speusippus, *Étude sur la doctrine pythagoricienne de la tétrade*, cited in

Kucharski, p. 21.

12. For the text of the oath and a commentary, see Delatte, "La tétractys pythagoricienne," p. 250.

13. Boethius, *De Musica*, book 1, chapter 10, *Fundamentals of Music*, p. 18.

CHAPTER TWO: OF MEASURED MULTITUDE

1. Plato, *Republic* 531c.

2. Aristotle, *Metaphysics* A.985b24–30.

3. Iamblichus, *Comm. Math.* 78.8–18, translation in Carl Huffman, *Philolaus of Croton: Pythagorean and Presocratic, A Commentary on the Fragments and Testimonia with Interpretative Essays* (Cambridge: Cambridge University Press, 1993), p. 70. On the Aristotelian authorship of the fragment, see Burkert, *Lore and Science*, pp. 49–50, n. 112.

4. Lives of Pythagoras by Diogenes, Porphyry, and Iamblichus survive. For a summary of the literature on Pythagoras's life, see Charles H. Kahn, *Pythagoras and the Pythagoreans: A Brief History* (Indianapolis: Hackett, 2001), pp. 5–22.

5. See Hermann Diels and Walter Kranz, *Die Fragmente der Vorsokratiker*, 7th ed. (Berlin: Weidmann, 1954), p. 18.

6. This is the so-called "Fourth Fragment" (Stob. Ecl. 1.21.7b): καὶ πάντα γα μὰν τὰ γιγνωσκόμενα ἀριθμὸν ἔχοντι. οὐ γὰρ οἷόν τε οὐδὲν οὔτε νοηθῆμεν οὔτε γνωσθῆμεν ἄνευ τούτου.

7. On early Greek conceptions of harmony, see Anne Gabrièle Wersinger's rich study, *La sphère et l'intervalle: Le schème de l'Harmonie dans la pensée des anciens Grecs d'Homère à Platon* (Paris: Jerôme Millon, 2008). On harmony in Heraclitus, see the helpful remarks in Charles H. Kahn, *The Art and Thought of Heraclitus: An Edition of the Fragments with Translation and Commentary* (Cambridge: Cambridge University Press, 1979), pp. 195–200. On the semantic field of the term ἁρμονία, cf. P. Bonaventura Meyer, *APMONIA: Bedeutungsgeschichte des Worte von Homer bis Aristoteles* (Zurich: A.-G. Gebr. Leemann & Co., 1925).

8. Kahn, *Pythagoras and the Pythagoreans*, p. 25.

9. Fragment 1 (Diog. 8.85): ἁ φύσις δ' ἐν τῶι κόσμωι ἁρμόχθη ἐξ ἀπείρων τε καὶ περαινόντων, καὶ ὅλος <ὁ> κόσμος καὶ τὰ ἐν αὐτῶι πάντα.

10. See the so-called "Fragment 1" (Porphyry in Ptolem. Harm., p. 56), reproduced and translated with a detailed commentary in Carl A. Huffman, *Archytas of Tarentum: Pythagorean, Philosopher, Mathematician King* (Cambridge: Cambridge

University Press, 2005), pp. 103–61. The evidence for Archytas's friendship with Plato is to be found in Plato's Seventh Letter, whose authorship has been disputed. It has been observed that Plato alludes to a "sorority" between the mathematical disciplines in *Republic* 530d.

11. Plato, *Epinomis*, 991e3–992b3.

12. For the Latin text, see Boethius, *Institution arithmétique*, ed. Jean-Yves Guillaumin (Paris: Les Belles Lettres, 1995), book 1, chapter 1, p. 6; English in Michael Masi, *Boethian Number Theory: A Translation of the De institutione arithmetica* (Amsterdam: Rodopi, 1993), p. 71.

13. Boethius, *Institution arithmétique*, book 1, chapter 1, paragraph 3, pp. 6–7; English in Masi, *Boethian Number Theory*, p. 72.

14. *Ibid.*

15. *Ibid.*

16. On the idea of a cycle of learning, see Ernst Robert Curtius, "Literature and Education," in *European Literature and the Latin Middle Ages*, trans. Willard R. Trask (Princeton: Princeton University Press, 1953), pp. 36–61.

17. The term does not appear before Boethius, though his classification is traditional. See Jean-Yves Guillaumin, "Le terme quadrivium de Boèce et ses aspects moraux," *L'Antiquité Classique* 59 (1990), pp. 139–48.

18. Plato, *Gorgias*, 451a–c. On Plato's distinction between arithmetic and logistic, see Jacob Klein, *Greek Mathematical Thought and the Origins of Algebra*, trans. Eva Brann (Cambridge, MA: The MIT Press, 1968), pp. 17–25.

19. Gottfried Friedlein, *Procli Diadochi in primum Euclidis Elementorum librum commentarii* (Leipzig: B. G. Teubner, 1873), p. 40, ll. 2–5; English in Klein, *Greek Mathematical Thought and the Origins of Algebra*, p. 12.

20. Karl Friedrich Hermann, *Platonis Dialogi* (Leipzig: B. G. Teubner, 1927), vol. 6, p. 290; English in Klein, *Greek Mathematical Thought and the Origins of Algebra*, p. 12.

21. Greek in Albert Jahn, "Olympiodori Philosophi Scholia in Platonis Gorgiam," *Neue Jahrbücher für Philologie und Pädagogik, oder kritische Bibliothek für das Schul- und Unterrichtswesen* 14.1 (1848), pp. 104–49, p. 131; English in Klein, *Greek Mathematical Thought and the Origins of Algebra*, p. 13.

22. Aristotle, *Physics* 4.12.220a27; *Metaphysics* 1.6.1056 b25ff.; Ð9.1085b10.

23. Euclid, *Elements*, book 7, definition 2: ἀριθμὸς δὲ τὸ ἐκ μονάδων συγκείμενον πλῆθος. That definition, in turn, presupposes the definition of the "unit" (book 7,

definition 1) as "that by virtue of which each of the things that exist is called one" (μονάς ἐστιν, καθ' ἥν ἔκαστον τῶν ὄντων ἓν).

24. That *arithmoi* are never anything other than "definite numbers of definite things" is the principal thesis of Klein's book; on the Greek notion of number in general, see *Greek Mathematical Thought and the Origins of Algebra*, chapter 6 ("The Concept of *Arithmos*," pp. 46–50). On the divergence in meaning between the classical *numerus* and the modern "number," see in particular p. 63, where Klein writes that "strictly speaking, it is not possible to call *arithmoi* 'numbers.'"

25. See Alexandre Koyré, "Du monde de l' 'à-peu-près' à l'univers de la précision," in *Études d'histoire de la pensée philosophique* (Paris: Armand Colin, 1961), pp. 311–29, esp. 313.

26. On the affinities between astronomy and harmonics, see Plato, *Republic* 531a–531e.

27. Greek in *Musici scriptores graeci: Aristoteles, Euclides, Nicomachus, Bacchus, Gaudentius, Alypius et melodiarum veterum quidquid extat*, ed. Karl von Jan (Leipzig: B. G. Teubner, 1895), p. 246, ll. 11–14: τὴν δὲ μεταξύτητα τῆς τε διὰ τεσσάρων καὶ τῆς διὰ πέντε ἀσύμφωνον μὲν ἑώρα αὐτὴν καθ' ἑαυτήν, συμπληρωτικὴν δὲ ἄλλως τῆς ἐν αὐτοῖς μειζονότητος; English in Andrew Barker (ed. and trans.), *Greek Musical Writings*, vol. 2, *Harmonic and Acoustic Theory* (Cambridge: Cambridge University Press, 1989), p. 256; *The Manual of Harmonics of Nicomachus the Pythagorean*, ed. and trans. Flora R. Levin (Grand Rapids, FL: Phanes Press, 1994), p. 83.

28. Greek in *Theonis Smyrnaei, philosophi platonici: Expositio rerum mathematicarum ad legendum Platonem utilium*, ed. Eduard Hiller (Leipzig: B. G. Teubner, 1878), p. 49, ll. 4–5: διάφωνοι δ' εἰσὶ καὶ οὐ σύμφωνοι φθόγγοι, ὧν ἐστι τὸ διάστημα τόνου.... ὁ γὰρ τόνος ... ἀρχὴ μὲν συμφωνίας, οὔπω δὲ συμφωνία.; English in Barker, *Greek Musical Writings*, vol. 2, *Harmonic and Acoustic Theory*, p. 213.

29. Greek in *In Nicomachi arithmeticam introductionem liber*, ed. H. Pistelli (Leipzig: B. G. Teubner, 1894), p. 11, ll. 1–2: μονὰς δέ ἐστι ποσοῦ τὸ ἐλάχιστον ἢ ποσοῦ τὸ πρῶτον καὶ κοινὸν μέρος ἢ ἀρχὴ ποσοῦ.

CHAPTER THREE: REMAINDERS

1. See *Elementa harmonica Aristoxeni*, ed. and trans. Rosetta da Rios (Rome: Typis Publicae Officinae Polygraphicae, 1954), esp. book 2, paragraph 44; English in Barker, *Greek Musical Writings*, vol. 2, *Harmonic and Acoustic Theory*, p. 159. On Aristoxenus's intervals, see R. P. Winnington-Ingram, "Aristoxenus and the

Intervals of Greek Music," *The Classical Quarterly* 26.3–4 (1932), pp. 195–208; for an overview of Greek scales, see M. L. West, "Scales and Modes," in *Ancient Greek Music* (Oxford: Oxford University Press, 1992), pp. 160–89.

2. For "syllable" as a name for the fourth, see Philolaus's so-called "Fragment 44," in Diels and Kranz, *Fragmente der Vorsokratiker*; cf. Theophrastus, Fragment 717, in *Theophrastus of Eresus: Sources for His Life, Writings, Thought and Influence,* ed. and trans. William W. Fortenbaugh, Pamela Huby, Robert W. Sharples, and Dimitri Gutas (Leiden: E. J. Brill, 1992), vol. 2, p. 573. On Philolaus's fragment, see Burkert, *Lore and Science in Ancient Pythagoreanism*, pp. 389–94; Huffman, *Philoloaus of Croton*, pp. 145–65; Andrew Barker, *The Science of Harmonics in Classical Greece* (Cambridge: Cambridge University Press, 2007), pp. 263–78. On the relation between musical and grammatical terminology, see Johannes Lohmann, *Mousiké und Logos: Aufsätze zur griechischen Philosophie und Musiktheorie, zum 75. Geburtstag des Verfassers am 9 Juli 1970,* ed. Anastasios Giannarás (Stuttgart: Musik-wissenschaftliche Verlagsgesellschaft, 1970), and on συλλαβή, see *ibid.*, pp. 9–11 and pp. 93–96.

3. Rios, *Elementa harmonic Aristoxeni*, 2.46; English in Barker, *Greek Musical Writings*, vol. 2, *Harmonic and Acoustic Theory*, p. 160.

4. It is not easy to reconstruct the teachings of the "harmonists" on the basis of the ancient sources. For a summary and analysis of the classical documentation, see Andrew Barker, *The Science of Harmonics in Classical Greece*, pp. 37–67.

5. See C. André Barbera, "Arithmetic and Geometric Divisions of the Tetra-chord," *Journal of Music Theory* 21.2 (1977), pp. 294–323.

6. "In the terminology of modern mathematics, the intervals, thought of as lengths of a line, correspond to the logarithms of the respective ratios." Burkert, *Lore and Science in Ancient Pythagoreanism*, pp. 369–70. For a succinct account of the resulting logarithmic correspondences, see Fabio Bellissima, "Epimoric Ratios and Greek Musical Theory," in Maria Luisa dalla Chiara, Roberto Giuntini, and Federico Laudisa (eds.), *Language, Quantum, Music: Selected Contributed Papers of the Tenth International Congress of Logic, Methodology and Philosophy of Science, Florence, August 1995* (Dordrecht: Kluwer, 1990), pp. 303–26, p. 305. For an account of the logistical procedure employed by the ancients for the calculation of these relations, see Theon of Smyrna, *Theonis Smyrnaei, Philosophi platonici: Expositio rerum mathematicarum ad legendum Platonem utilium,* ed. Eduard Hiller (Leipzig: B. G. Teubner, 1878), p. 67, line 16 through p. 72, line 70.

7. Boethius, *De musica*, book 1, chapter 3; Latin in Meyer, *Traité de musique*, p. 38.

8. In the terms of classical arithmetic, the ratio defining the octave (2:1) is technically "multiple," since it consists of two quantities, "one of which contains the other two, three, four or more times, without subtracting anything from it or adding anything to it." Boethius, *De musica*, book 1, chapter 4. To this degree, its structure differs from that of the other Pythagorean consonances, yet it remains no less incapable of being "halved."

9. This thesis can also be expressed by stating that there is no geometrical mean for two numbers in superparticular proportion.

10. See *Sectio canonis*, proposition 3, in André Barbera, *The Euclidean Division of the Canon: Greek and Latin Sources* (Lincoln: University of Nebraska Press, 1991), pp. 194–97; cf. Thomas J. Mathiesen, "An Annotated Translation of Euclid's 'Division of the Monochord,'" *Journal of Music Theory* 19.2 (1975), p. 240; Barker, *Greek Musical Writings*, vol. 2, *Harmonic and Acoustic Theory*, p. 195. The Euclidean authorship of this text has been contested, most famously by Tannery: see Paul Tannery, "Un traité grec d'arithmétique antérieur à Euclide," in J. L. Heiberg and H. G. Zeuthen (eds.), *Mémoires scientifiques*, 7 vols. (Paris: E. Privat, 1912–25), vol. 7, pp. 244–50.

11. See Boethius, *Traité de la musique*, book 4, chapter 2, p. 226: "Superparticularis intervalli medius numerus neque unus neque plures proprotionaliter intervenient." Boethius's discussion of the theorem can be found in book 3, chapters 1–4. For a comparison of the Latin and Greek versions of the principle, see Tannery, "Un traité grec d'arithmétique antérieur à Euclide," pp. 244–50; B. L. van der Waerden, "Arithmetik der Pythagoreer," *Mathematische Annalen* 120 (1947–49), pp. 127–53 and 676–700, esp. 132–36; Burkert, *Lore and Science in Ancient Pythagoreanism*, pp. 442–47. For an analysis of Archytas's theorem, see Wilbur Richard Knorr, *The Evolution of the Euclidean Elements: A Study of the Theory of Incommensurable Magnitudes and Its Significance for Early Greek Geometry* (Dordrecht: Kluwer, 1975), pp. 212–24.

12. Archytas and Plato were certainly familiar with this operation, which may date back to Philolaus. For discussions, see Burkert, *Lore and Science in Ancient Pythagoreanism*, pp. 388–89; Huffman, *Archytas of Tarentum*, pp. 420–22; Barker, *The Science of Harmonics in Classical Greece*, pp. 269–73.

13. *Timaeus* 36b. On the harmonic structure of the Platonic world soul, see

August Boeckh, "Über die Bildung der Weltseele im Timaeos des Platon," in *Gesammelte Kleine Schriften*, vol. 3: *Reden und Abhandlungen*, ed. Ferdinand Ascherson (Leipzig: B. G. Teubner, 1866), pp. 109–80; Evanghélos Moutsopoulos, *La musique dans l'oeuvre de Platon* (Paris: Presses universitaires de France, 1959), pp. 249–54; Sergio Zedda, "How to Build a World Soul: A Practical Guide," in *Reason and Necessity: Essays on Plato's Timaeus*, ed. M. R. Wright (London: Duckworth, 2000), pp. 23–41; Baker, *The Science of Harmonics in Classical Greece*, pp. 318–23.

14. Pythagoreans also referred to the interval as δίεσις: see Theo. Sm. 55.11; Chalcid. 45; Macrob. *Somn. Sc.* 2.1.23; Proclus, *In Tim.* 2.168.28; Boethius, *De musica*, book 2, chapter 28.

15. According to Boethius, Philolaus went even further, bisecting the "comma." That division, of course, violates Archytas's theorem, since "if the comma can be bisected," Burkert notes, "surely a whole tone may." (*Lore and Science in Ancient Pythagoreanism*, p. 398.) On the alleged division of the *diesis* into two "diaschisms" and the comma into two "schisms," cf. also Paul Tannery, "À propos des fragments Philolaïques sur la musique," *Mémoires scientifiques*, vol. 3, pp. 220–43 and 223–25.

16. Burkert, *Lore and Science in Ancient Pythagoreanism*, p. 395.

17 Rios, *Elementa Harmonica Aristoxeni*, book 2.4; English in Barker, *Greek Musical Writings*, vol. 2, *Harmonic and Acoustic Theory*, p. 135.

18. Hiller, *Theonis Smyrnaei, Philosophi platonici: Expositio rerum mathematicarum ad legendum Platonem utilium*, p. 18, ll. 18–21. On fractions in the Greek logistics, see Klein, *Greek Mathematical Thought and the Origins of Algebra*, pp. 37–46.

19. For a comprehensive treatment of ancient tunings in modern terms, see J. Murray Barbour, *Tuning and Temperament: A Historical Survey* (East Lansing: Michigan State University Press, 1953), pp. 15–24. Cf. also J. Murray Barbour, "The Persistence of the Pythagorean Tuning System," *Scripta Mathematica* 1 (1933), pp. 286–304.

CHAPTER FOUR: DISPROPORTIONS

1. Greek in *Iamblichi De vita Pythagorica Liber*, ed. Ludwig Deubner (Leipzig: G. B. Teuber, 1937), 246–47, p. 132.

2. *Ibid.*, 247, p. 132.

3. Arabic in William Thomson and Gustav Junge (eds.), *The Commentary of Pappus on Book X of Euclid's Elements: Arabic Text and Translation* (Cambridge, MA: Harvard University Press, 1930), book 1, sections 1–2, pp. 191–92; English in *ibid.*,

pp. 63–64.

4. The literature on the discovery of incommensurability is vast. In addition to the works cited in the notes to this chapter, see Hieronymus Georg Zeuthen, "Sur l'origine historique de la connaissance des quantités irrationelles," *Oversigt over det Koneglige Danske videnskabernes Selskabs Ferhandlinger* (1915), pp. 333–52; Heinrich Scholz, "Warum haben die Griechen die Irrationalzahlen nicht aufgebaut?" *Kant-Studien* 33 (1928), pp. 35–72; Heinrich Vogt, "Die Entdeckungsgeschichte des Irrationalen nach Plato und anderen Quellen des 4. Jahrhundert," *Bibliotheca mathematica* 10 (1909–1910), pp. 97–155; G. Junge, "Wann haben die Griechen das Irrationale entdeckt?" in *Novae Symbolae Joachimicae: Festschrift des königlichen Joachimsthalschen Gymnasium* (Halle: Verlag der Buchhandlung des Waisenhauses, 1907), pp. 221–66.

5. Knorr, *The Evolution of the Euclidean Elements*, p. 29.

6. Two lines are in mean and extreme ratio if the larger is the mean proportion between the smaller and their sum. Where A is the longer line and B the shorter, $A:B = B:A + B$. For an overview of Greek geometrical treatments of the golden mean, see François Lasserre, *La naissance des mathématiques à l'époque de Platon* (Paris: Éditions du Cerf, 1990), pp. 150–57.

7. Euclid, *Elements*, book 10, proposition 2; English in *The Thirteen Books of Euclid's Elements*, vol. 3, *Books X–XIII and Appendix*, trans. Sir Thomas Heath, 2nd ed., 3 vols. (Cambridge: Cambridge University Press, 1956), p. 17. On reciprocal subtraction as the origin of the discovery of incommensurability, see Kurt von Fritz, "The Discovery of Incommensurability by Hippasus of Metapontum," in David J. Furley and R. E. Allen (eds.), *Studies in Presocratic Philosophy*, 2 vols. (New York: Humanities Press, 1970–1975), vol. 1, pp. 409–12, and Siegfried Heller, "Die Entdeckung der stetige Teilung durch die Pythagoreer," *Abhandlungen der Deutschen Akademie der Wissenschaften zu Berlin, Klasse für Mathematik, Physik und Technik* 6 (1958), pp. 9–11. Cf. Knorr, *The Evolution of the Euclidean Elements*, pp. 29–30.

8. On the theorem in Greek mathematics, see Sir Thomas Heath, *A History of Greek Mathematics*, 3 vols. (Oxford: Clarendon press, 1921), vol. 2, pp. 144–54.

9. See *Theaetetus* 147c–148d; *Meno* 82b–85b.

10. The proof is offered as an example of reasoning *per absurdum*. See *Prior Analytics*, 1.23.41a29.

11. See Lasserre, *La naissance des mathématiques à l'époque de Platon*, p. 93. For the Euclidean formulation, see the so-called "proposition 117" of book 10. Cf.

Knorr, *The Development of the Euclidean Elements*, pp. 22–28.

12. It may be recalled that in the *Gorgias*, arithmetic is defined as knowledge of "the even and the odd, with reference to how much either happens to be" (451a–c). On the development of the Greek theory of the even and the odd, see Oskar Becker, "Die Lehre vom Geraden und Ungeraden im Neunten Buch der Euklidischen Elemente: Versuch einer Wiederherstellung in der urpsrünglichen Gestalt," *Quellen und Studien zur Geschichte der Mathematik, Astronomie und Physik B, Studien* 3 (1936), pp. 533–53.

13. Simone Weil, *Sur la science* (Paris: Gallimard, 1966), p. 217.

14. See Tannery, "Du rôle de la musique grecque dans le développement de la mathématique pure," *Mémoires scientifiques*, vol. 3, pp. 68–89. On music and the discovery of the incommensurable, see also the various studies by Árpád Szabó collected in *Anfänge der griechischen Mathematik* (Budapest: Akadémiai Kiádo, 1969) and *Die Entfaltung der griechischen Mathematik* (Leipzig: Bibliographisches Institut, 1994). On the role of the irrational in the Pythagoreans' foundation of Western musical theory, see Hugues Dufourt's far-reaching *Essai sur les principes de la musique, vol. 1, Mathêsis et subjectivité: Des conditions historiques de possibilité de la musique occidentale* (Paris: Musica Ficta, 2007).

15. Pythagorean mathematics had defined "proportions" as the relations of four terms in which "the first is the same multiple, or the same part, or the same parts, of the second that the third is of the fourth" (as in the definition of proportional numbers in *Elements*, book 7, definition 20). In such a theory, incommensurable magnitudes would be excluded from the field of proportion, lacking by definition any "integral multiples" or aliquot parts among themselves. Eudoxus offered a new definition of proportion that accommodated incommensurable magnitudes: "Magnitudes are said to be in the same ratio, the first to the second and the third to the fourth, when, if any equimultiples whatever be taken of the first and third, and any equimultiples whatever of the second and the fourth, the former multiples also exceed, are alike to, or alike fall short of, the latter equimultiples respectively taken in corresponding order" (*Elements*, book 5, definition 5). One may formulate the principle symbolically as "$A/B = C/D$ if, and only if, for all positive integers, m, n, when $nA \gtrless mB$ then, correspondingly $nC \gtrless mD$." *Elements*, book 5, definition 4 stipulates the condition that magnitudes must satisfy to be considered homogenous. For a summary and account of the fate of these principles after antiquity, see John E. Murdoch, "The Medieval Language of Proportions: Elements of the

Interaction with Greek Foundations and the Development of New Mathematical Techniques," in A. C. Crombie, *Scientific Change: Historical Studies in the Intellectual, Social and Technical Conditions for Scientific Discovery and Technical Invention, from Antiquity to the Present* 1 (New York: Basic Books, 1963), pp. 238–39. On Eudoxus's doctrine of proportions, see Lasserre, *La naissance des mathématiques à l'époque de Platon*, pp. 127–78; Jean-Louis Gardiès, *L'héritage épistémologique d'Eudoxe de Cnide* (Paris: Vrin, 1968).

16. Marie-Elisabeth Duchez, "Des neumes à la portée: Élaboration et organization rationnelles de la discontinuité musicale et sa representation graphique, de la formule mélodique à l'échelle monocordale," *Revue de musique des universités canadiennes* 4 (1983), pp. 39–40. The literature on the development of neumatic notation is abundant. For a summary, see Kenneth Levy, "On the Origin of Neumes," *Early Music History* 7 (1987), pp. 59–90.

17. Duchez, "Des neumes à la portée," p. 55.

18. See Willi Apel, "Mathematics and Music in the Middle-Ages," in *Medieval Music: Collected Articles and Reviews*, forward by Thomas E. Binkley (Stuttgart: Franz Steiner Verlag, 1986), p. 144.

19. On the Church modes, see David E. Cohen, "Notes, Scales, and Modes in the Earlier Middle Ages," in Thomas Christensen (ed.), *The Cambridge History of Western Music Theory* (Cambridge: Cambridge University Press, 2002), pp. 307–63, esp. 309–13.

20. For a summary of John of Garland's theory of orders, see Anna Maria Busse Berger, "The Evolution of Rhythmic Notation," in *The Cambridge History of Western Music Theory*, pp. 628–31.

21. See Willi Apel, *The Notation of Polyphonic Music*, 5th ed. (Cambridge, MA: The Medieval Academy of America, 1953), p. 145.

22. For an overview of proportions, see *ibid.*, pp. 145–79; cf. Berger, "The Evolution of Rhythmic Notation," in *The Cambridge History of Western Music Theory*, pp. 628–56.

23. Apel, *The Notation of Polyphonic Music*, p. 145.

24. *Ibid.*, pp. 145–46.

25. *Ibid.*, p. 147.

26. See *Nicole Oresme and the Medieval Geometry of Qualities and Motions: A Treatise on the Uniformity and Difformity of Intensities Known as Tractatus de configurationibus qualitatum et motuum*, ed. and trans. Marshall Clagett (Madison:

The University of Wisconsin Press, 1968). Clagett dates the treatise to the period between 1351 and 1355. See "Introduction," p. 125.

27. Clagett, *Tractatus de configurationibus qualitatum et motuum*, book 1, part 1, pp. 164–65.

28. See Roger Bacon's *De graduatione medicinarum compositarum,* in *Opera hactenus inedita,* ed. Robert Steele (Oxford: Clarendon Press, 1905–), 16 vols., fasc. 9, pp. 144–49. For an analysis of its relation to Oresme, see Anneliese Maier, *Zwei Grundprobeme der scholastischen Naturphilosophie: Das Problem der intensiven Grüße; Die Impetustheorie,* 3rd ed. (Rome: Edizioni di Storia e letteratura, 1968), pp. 97–98. For a discussion of the history of such representations, see Clagett, *Nicole Oresme and the Medieval Geometry of Qualities and Motions,* pp. 50–73.

29. On the medieval theory of "latitude," see Anneliese Maier, *An der Grenzen von Scholastik und Naturwissenschaft: Die Struktur der materiellen Substanz, Das Problem der Gravitation, Die Mathematik der Formlatituden,* 2nd ed. (Rome: Storia e letteratura, 1952), pp. 257–384; Edith D. Sylla, "Medieval Concepts of the Latitude of Forms: The Oxford Calculators," *Archives d'histoire doctrinale et littéraire du Moyen Âge* 40 (1973), pp. 223–83.

30. Clagett, *Tractatus de configurationibus qualitatum et motuum,* book 1, part 4, pp. 172–76.

31. *Ibid.,* book 1, part 1, pp. 166–67.

32. See Sylla, "Medieval Concepts of the Latitude of Forms," esp. pp. 226–33; Michael McVaugh, "Arnald of Villanova and Bradwardine's Law," *Isis* 58.1 (1967), pp. 56–64.

33. The discussion of musical phenomena runs from book 2, part 15, to book 2, part 24. On this section of the *Tractatus de configurationibus,* see V. Zoubov, "Nicole Oresme et la musique," *Medieval and Renaissance Studies* 5 (1961), pp. 96–107; Fabrizio della Seta, "Idee musicali nel *Tractatus de configurationibus qualitatum et motuum* di Nicola Oresme," in *La musica nel tempo di Dante: Atti del convegno internazionale, Ravenna, 12–14 Settembre 1986* (Milan: Unicopli, 1988), pp. 222–56; Ulrich Taschow, "Die Bedeutung der Musik als Modell für Nicole Oresmes Theorie: *De configurationibus qualitatum et motuum,*" *Early Science and Medecine* 4.1 (1999), pp. 37–90.

34. Clagett, *Tractatus de configurationibus qualitatum et motuum,* book 2, part 15, pp. 306–307.

35. *Ibid.*

36. For Oresme's account of instrumental timbres, see Clagett, *Tractatus de configurationibus qualitatum et motuum*, book 2, part 18. Cf. Zoubov, "Nicole Oresme et la musique," p. 101.

37. *Nicole Oresme and the Kinematics of Circular Motion: Tractatus de commensurabilitate vel incommensurabilitate motuum celi*, ed and trans. Edward Grant (Madison: The University of Wisconsin Press, 1971).

38. See part 3 of Grant, "Introduction," in *Nicole Oresme and the Kinematics of Circular Motion*, pp. 78–161 and, on Theodosius especially, pp. 78–86. The Greek text of Theodosius's treatise was edited and translated into Latin by Rudolf Fecht, *Theodosii De habitationibus; De diebus et noctibus, Abhandlungen der Gesellschaft der Wissenschaften zu Göttingen*, phil.-hist. Klasse, n.s. (Berlin: 1927), vol. 19, p. 4.

39. On Oresme and Johannes de Muris, see Grant, "Introduction," pp. 86–161.

40. Where V is velocity, F is force or motive power, and R is the resistance of medium or mobile, then $V \propto F/R$.

41. Aristotle nowhere explicitly denies the possibility of such an operation, though the medieval principle that force must be greater than resistance for there to be speed could be derived from *Physics* 7.5.250a15–20. Even if one grants that motive force must exceed the resistance of a mobile or medium, difficulties will arise. See Grant's remarks in the "Introduction" to *Nicole Oresme, De proportionibus proportionum and Ad pauca respicientes*, ed. and trans. Edward Grant (Madison: The University of Wisconsin Press, 1966), pp. 16–17.

42. See Marshall Clagett, *Giovanni Marliani and Late Medieval Physics* (New York: Columbia University Press, 1941), p. 129.

43. The expression was coined by Anneliese Maier. See her pioneering study, "Die Funktionsbegriff in der Physik des 14. Jahrhundert," in *Die Vorläufer Galileis im 14. Jahrhundert*, 2nd ed. (Rome: Storia e letteratura, 1966), pp. 81–110. For Bradwardine's treatise, see *Tractatus proportionum seu de proportionibus velocitatum in motibus*, in *Thomas of Bradwardine: His Tractatus de Proportionibus*, ed. H. Lamar Crosby, Jr. (Madison: The University of Wisconsin Press, 1955).

44. One may also express the rule by stating "the proportion of velocities in motions follows the proportion of the power of the motor to the power of the thing moved." See Marshall Clagett, *The Science of Mechanics in the Middle Ages* (Madison: The University of Wisconsin Press, 1959), p. 438. In more modern terms, the "dynamical law" can be expressed as "$F_2/R_2 = (F_1/R_1)^n$ where $F_1/R_1 > 1$ and $n = V_2/V_1$," or, more succinctly, "$V = \log_a (F/R)$; where $a = F_1/R_1$." On the

limitations of such "translations," see Alexandre Koyré's review of Anneliese
Maier, *Die Vorläufer Galileis im 14. Jahrhundert: Studien zur Naturphilosophie der
Spätscholastik* (Rome: Editizioni Storia e letteratura, 1949), *Archives internationales d'histoire des sciences*, n. s. d'Archeion 4.14 (30) (1951), pp. 769–83, esp.
775–76. Stillman Drake has argued that in relating force to resistance, Bradwardine himself, unlike his followers, did not envisage fractional exponents. See
"Bradwardine's Function, Mediate Denomination and Multiple Continua," *Physis*
12 (1970), pp. 51–68.

45. The principle constitutes the tenth proposition of book 3 of *De Proportionibus proportionum*: "It is probable that two proposed unknown ratios are incommensurable because if many unknown ratios are proposed it is most probable that
any [one] would be incommensurable to any [other]." See Grant, *Nicole Oresme,
De proportionibus proportionum and Ad pauca respicientes*, pp. 246–47 and, for the
demonstration, pp. 246–55. The rule can also be formulated as follows: "if A/B and
C/D are two proportions randomly chosen from the given set, then it is probable
(*verisimile*) that $A/B \neq (C/D)^{m/n}$ — where m and n are positive integers." Murdoch,
"The Medieval Language of Proportions," p. 268.

46. See Grant, *Nicole Oresme, De proportionibus proportionum and Ad pauca
respicientes*, book 4, proposition 7, pp. 299–309, esp. 304–305.

47. Grant, *Nicole Oresme and the Kinematics of Circular Motion*, pp. 288–89.

48. *Ibid.*, pp. 294–95; the allusion is to *De institutione arithmetica*, 1.2.14–15,
though the original differs considerably from this citation, as Grant indicates in
his note, p. 341 n. 20.

49. *Ibid.*, pp. 294–95; the reference is to the *Commentary on the Dream of Scipio*,
ed. William Harris Stahl (New York: Columbia University Press, 1952), 1.5.12;
Latin in *Ambrosii Theodosii Macrobii Commentarii in somnium Scipionis*, ed. Jacob
Willis (Leipzig: G.B. Teubner, 1963).

50. Grant, *Nicole Oresme and the Kinematics of Circular Motion*, pp. 294–95.

51. *Ibid.*, pp. 130–31. Arithmetic draws her words from the Pseudo-Aristotelian
De mundo (6, 397b29–34), as Grant indicates (see pp. 341–42 n. 23).

52. Grant, *Nicole Oresme and the Kinematics of Circular Motion*, pp. 294–95.

53. *Ibid.*, pp. 296–97.

54. *Ibid.*

55. *Ibid.*, pp. 312–13.

56. *Ibid.*, pp. 314–15.

57. *Ibid.*

58. *Ibid.*, Grant, *Nicole Oresme and the Kinematics of Circular Motion*, pp. 316–17.

59. *Ibid.*

60. *Ibid.*, pp. 321–22.

61. Nicole Oresme, *Le livre du ciel et du monde*, ed. Albert D. Menut and Alexander J. Denomy, trans. Albert D. Menut (Madison: The University of Wisconsin Press, 1968), p. 196. The passage was first noted by Zoubov: see "Nicole Oresme et la musique," pp. 102–103. Cf. Grant, "Introduction," *Nicole Oresme and the Kinematics of Circular Motion*, pp. 72–77.

62. Grant, *Nicole Oresme and the Kinematics of Circular Motion*, pp. 322–23

CHAPTER FIVE: CIPHERS

1. Robert W. Wienpahl, "Zarlino, the Senario, and Tonality," *Journal of the American Musicological Society*, 12.1 (1959), pp. 27–41, p. 27.

2. Gioseffe Zarlino, *Instituzioni armoniche* (Venice: n.p., 1558), book 1, chapter 12, p. 21.

3. *Ibid.*

4. *Ibid.*, p. 22.

5. *Ibid.*

6. For the introduction of the term *senario*, see *ibid.*, book 1, chapter 14, pp. 23–24.

7. *Ibid.*, p. 24.

8. H. F. Cohen, *Quantifying Music: The Science of Music at the First Stage of the Scientific Revolution, 1580–1650* (Dordrecht: D. Reidel, 1984), p. 7.

9. See Cohen's remarks in *ibid.*, p. 6.

10. On the intervals whose relations contain the number 7, see Cohen, *Quantifying Music*, pp. 6, 107–11, 225–28.

11. See Franchinus Gaffurius, *Theorica musicae*, ed. Ilde Illuminati and Cesarino Ruini, trans. Ilde Illuminati, with an essay by Fabio Bellissima (Florence: Edizioni del Galluzzo, 2005), book 1, chapter 8, pp. 67–71.

12. Vincenzo Galilei, *Discorso intorno alle opere di Gioseffo Zarlino et altri importanti particolari attenenti alla musica* (Venice: n.p., 1589), pp. 103–104. A translation of this passage can be found in Claude V. Palisca, "Scientific Empiricism in Musical Thought," in Hedley Howell Rhys (ed.), *Seventeenth Century Science and the Arts* (Princeton: Princeton University Press, 1961), p. 128.

13. The point was to be clearly presented by Galilei's son. See *Discorsi sopra due nuove scienze* (1638), in Galileo Galilei, *Opere*, ed. Franz Brunetti, 2 vols. (Turin: UTET, 1964), vol. 2, p. 673.

14. See Palisca, "Scientific Empiricism in Musical Thought," p. 128.

15. The arithmetical inequalities that define the major and minor thirds (5:4 and 6:5) are superparticular in form. The sixths appear not to be, but Zarlino argued that the major sixth (5:3) and the minor sixth (8:5) could be regarded as the union of a perfect fourth (4:3) and a major (5:4) or minor (6:5) third.

16. For the *numerus sonorus* in Zarlino, see *Instituzioni armoniche*, book 1, chapter 19, pp. 29–30.

17. Vincenzo Galilei argued that even as linear string length corresponds to direct proportionality and two-dimensional mass corresponds to squared proportionality, the pitches of voluminous bodies, such as pipes, correspond to cubed proportionality. But the truth is that the pitch of a pipe relates to its length, not its volume. See D. P. Walker, "Some Aspects of the Musical Theory of Vincenzo Galilei and Galileo Galilei," *Proceedings of the Royal Musical Association* 100 (1973–1974), pp. 33–47 and 42–43; Cohen, *Quantifying Music*, pp. 82–83. On Zarlino, Vincenzo Galilei, and the impact of experimental science on early modern music theory, see also D. P. Walker, "Musical Humanism in the 16th and Early 17th Centuries," *The Music Review* 2 (1941), pp. 1–13; *The Music Review* 2.2, pp. 111–21; *The Music Review* 2.3 (1942), pp. 220–27; *The Music Review* 2.4 (1942), pp. 288–308; *The Music Review* 3 (1942), pp. 55–71; Stillman Drake, "Renaissance Music and Experimental Science," *Journal of the History of Ideas* 31.4 (1970), pp. 483–500.

18. Greek in *Alexandri Aphrodisiensis in Aristotelis Metaphysica commentaria*, ed. Michael Hayduck (Berlin: G. Reimer, 1891), 86.5; cited in Klein, *Greek Mathematical Thought and the Origins of Algebra*, p. 48.

19. Klein, *Greek Mathematical Thought and the Origins of Algebra*, p. 174.

20. *Ibid.*, p. 175.

21. *Ibid.*, pp. 192–93.

22. See Gérard Stevin, *Les oeuvres mathématiques de Simon Stevin de Bruges* (Leiden: n. p. 1634), cited in Klein, *Greek Mathematical Thought and the Origins of Algebra*, p. 193.

23. The literature on the zero is sizable. For an overview of some of the questions relating to the problem of its introduction into mathematical notation, see Carl B. Boyer, "Zero: The Symbol, the Concept, the Number," *National*

Mathematics Magazine 18.8 (1944), pp. 323–30.

24. See Simon Stevin, *L'arithmétique* (Leyden: Chrisophe Plantin, 1585), p. 3v, in Ernst Crone, E. J. Diksterhuis, R. J. Forbes, M. G. Minnaert, and A. Pannekoek (eds.), *The Principal Works of Simon Stevin*, 5 vols. in 6 (Amsterdam: Swets and Zeitlinger, 1955–), vol. 2, part B, *Mathematics*, ed. D. J. Struik, pp. 501 and 499, cited in Klein, *Greek Mathematical Thought and the Origins of Algebra*, p. 193. Wallis held a similar view: cf. *ibid.*, p. 214.

25. Stevin, *L'arithmétique*, p. 4v: *Nombre n'est poinct quantité discontinue*, in *The Principal Works of Simon Stevin*, vol. 2, part B: *Mathematics*, ed. Struik, p. 501, cited in Klein, *Greek Mathematical Thought and the Origins of Algebra*, p. 194.

26. *Ibid.*, p. 5r, in *The Principal Works of Simon Stevin*, vol. 2, part B: *Mathematics*, ed. Struik, p. 502, cited in Klein, *Greek Mathematical Thought and the Origins of Algebra*, p. 195.

27. John Wallis, *Mathesis universalis, sive, Arithmeticum opus integrum, tum Philologice, tum Mathematice traditum, Arithmeticam tum Numerosam, tum Speciosam sive Symbolicam complectens, sive Calculum Geometricum; tum etiam Rationum proportionumve traditionem; Logarithmorum item Doctrinam; aliaque, quae Capitum Syllabus indicabit*, chapter 35, in *Opera mathematica* (Oxoniae, e theatro Sheldoniano, 1693–1699), 3 vols., vol 1, p. 183, cited in Klein, *Greek Mathematical Thought and the Origins of Algebra*, p. 220.

28. Klein, *Greek Mathematical Thought and the Origins of Algebra*, p. 221.

29. On the difference between the ancient cosmos and the empirical and mathematicizable universe, see the classic work by Alexandre Koyré, *From the Closed World to the Infinite Universe* (Baltimore: Johns Hopkins University Press, 1957). For an axiomatic presentation of Koyré's theory of science as supplemented by Alexandre Kojève and employed by Jacques Lacan, see Jean-Claude Milner, *L'œuvre claire: Lacan, la science, la philosophie* (Paris: Seuil, 1995) and, on the distinction between the mathematical and the mathematizable in particular, pp. 52–53.

CHAPTER SIX: TEMPERAMENTS

1. Giovanni Battista Benedetti, *Diversarum speculationum mathematicarum et physicarum liber* (Taurini: apud Haeredem Bevilaquae, 1585).

2. *Ibid.*; in my translation, I have been aided by Claude V. Palisca's rendition: see his fundamental study, "Scientific Empiricism in Musical Thought," p. 107.

3. See *ibid.*, pp. 107–109.

4. See *ibid.*, p. 108: "Quidem numeri absque mirabili analogia conveniunt invicem."

5. See the remarks in H. F. Cohen, *Quantifying Music: The Science of Music at the First Stage of the Scientific Revolution, 1580–1650* (Dordrecht: D. Reidel, 1984), pp. 76–77.

6. For a summary, see Cohen, *Quantifying Music*, pp. 88–90; for a more sustained study presentation, see François Baskevitch, "L'élaboration de la notion de vibration sonore: Galilée dans les *Discorsi*," *Revue d'histoire des sciences* 60.2 (2007), pp. 387–418.

7. Similarly, Beeckman showed that the width or amplitude of a string is directly proportionate to the loudness of sound. See Cohen, *Quantifying Music*, pp. 124–27.

8. *Ibid.*, p. 120.

9. *Ibid.*, pp. 120–23.

10. Thus, it was an axiom that consonances result from epimoric (or "superparticular") and multiple ratios, as Euclid established in his *Sectio canonis*. See Chapter 5, "Ciphers."

11. Palisca, "Scientific Empiricism in Musical Thought," p. 109.

12. On the history and longevity of the Pythagorean tuning, see J. Murray Barbour, "The Persistence of the Pythagorean Tuning System."

13. For a helpful summary of incommensurable intervals and their consequences for tuning and temperament, see Mark Lindley, "Temperament," in *The New Grove Dictionary of Music and Musicians*, ed. Stanley Sadie, 20 vols. (London: Macmillan, 1980), pp. 660–74; J. Murray Barbour, *Tuning and Temperament*.

14. It might be expected that after a progression of twelve fifths, one would reach a tone that sounded in unison with the first tone in the progression, but this cannot be.

15. See Lindley, "Temperament," pp. 660–61.

16. Barbour, *Tuning and Temperament*, pp. 1–13; on Aristoxenus more generally, see Barker, *The Science of Harmonics in Classical Greece*, pp. 136–54.

17. Cited in Barbour, *Tuning and Temperament*, p. 25.

18. See Barbour, *Tuning and Temperament*; Patrice Bailhache, *Une histoire de l'acoustique musicale* (Paris: CNRS Éditions, 2001), pp. 95–98.

19. See Barbour, *Tuning and Temperament*, on irregular temperaments; on Andreas Werckmeister, see Bailhache's summary in *Une histoire de l'acoustique*

musicale, pp. 100–101. For Werckmeister's book, see *Musicalische Temperatur*, ed. Rudolf Rasch (Utrecht: Diapason Press, 1983).

20. See the summary in Barbour, *Tuning and Temperament*, pp. 6–7.

21. On Salinas's proposition for a meantone temperament, see Barbour, *Tuning and Temperament*, pp. 33–35.

22. On Huygens and multiple division temperaments, see Barbour, *Tuning and Temperament*, pp. 107–32, esp. 118–20; Bailhache, *Une histoire de l'acoustique musicale*, pp. 101–104.

23. See Bailhache, *Une histoire de l'acoustique musicale*, pp. 104–106.

24. Letter to Goldbach of 1712, in Gottfried Wilhelm Leibniz, *Epistolae ad diversos theologici, iuridici, medici, philosophici, mathematici, historici et philologici argumenti e msc. auctoris cum annotationibus suis primum divulgavit*, ed. Christian Kortholt, 2 vols. (Leipzig: 1734), vol 1, pp. 238–42. Cf. the similar terms in "De Rerum originatione radicali," in *Die philosophische Schriften von Gottfried Wilhelm Leibniz*, ed. C. J. Gerhardt, 7 vols. (1875–1890;. Hildesheim: Olms, 1978), vol. 7, pp. 302–308 and 306–307; English in Gottfried Wilhelm Leibniz, *Philosophical Essays*, trans. Roger Ariew and Daniel Garber (Indianapolis: Hackett, 1989), pp. 149–154 and 153.

25. Cohen, *Quantifying Music*, p. 200 n. 43; cf. Palisca, "Scientific Empiricism in Musical Thought," pp. 110–11.

26. Leibniz to Henfling, Hanover, Summer 1706, in *Der Briefwechsel zwischen Leibniz und Conrad Henfling*, ed. Rudolf Haase (Frankfurt am Main: Vittorio Klostermann, 1982), p. 58. On Leibniz's correspondence with Henfling, see Patrice Bailhache, "Le miroir de l'Harmonie Universelle: Musique et théorie de la musique chez Leibniz," in *L'esprit de la musique: Essais d'esthétique et de philosophie*, ed. Hugues Dufourt, Joël-Marie Fauquet, and François Hurard (Paris: Klincksieck, 1992), pp. 203–16, esp. 209–11. On Leibniz and music more generally, see Patrice Bailhache, *Leibniz et la théorie de la musique* (Paris: Klincksieck, 1992); Yvon Belaval, "L'idée d'harmonie chez Leibniz," in *Histoire de la philosophie: Ses problèmes, ses méthodes, Hommage à Martial Guéroult* (Paris: Fischbacher, 1964), pp. 59–78; Andrea Luppi, *Lo specchio dell'armonia universale: Estetica e musica in Leibniz* (Milan: Franco Angeli, 1989); Fabrizio Mondadori, "A Harmony of One's Own and Universal Harmony in Leibniz's Paris Writings," in *Leibniz à Paris (1672–1676): Symposium de la Leibniz-Gesellschaft (Hannover) et du Centre National de la Recherche Scientifique (Paris) à Chantilly (France) du 14 au 18 novembre, 1976, 2*

vols. (Wiesbaden: F. Steiner, 1978), vol. 2, *La Philosophie de Leibniz*, pp. 151–68. On Conrad Henfling, see Patrice Bailhache, "Le système musical de Conrad Henfling (1706)," *Revue de musicologie* 74.1 (1981), pp. 5–25.

27. Haase, *Der Briefwechsel zwischen Leibniz und Conrad Henfling*, p. 58.

28. See Gottfried Wilhelm Leibniz, "Principles of Nature and Grace Founded on Reason," paragraph 17, in *Die philosophische Schriften*, vol. 6, p. 605; English in *Philosophical Essays*, p. 212.

29. For the classic account of Leibniz's theory of the degrees of cogitation, see the "Meditationes de cognitione, veritate et ideis" of 1684: "Est ergo cognitio vel obscura vel clara, et clara rursus vel confusa vel distincta, et distincta vel inadaequata vel adaequata, item vel symbolica vel intuitiva: et quidem si simul adaequata et intuitiva sit, perfectissima est," in *Die philosophische Schriften*, vol. 4, p. 422.

30. Gottfried Wilhelm Leibniz, "Discours de Métaphyisque," in *Die philosophische Schriften*, vol. 4, p. 449.

31. For Baumgarten's introduction of the term "aesthetics," see Alexander Gottlieb Baumgarten, *Meditationes philosophicae de nonnullis ad poema pertinentibus*, ed. and trans. Heinz Paetzold (Hamburg: Felix Meiner, 1983), section 117, p. 86; for the doctrine of his aesthetics proper, see Alexander Gottlieb Baumgarten, *Aesthetica*, 3rd ed. (1750; Frankfurt an der Oder: Olms, 1986).]

CHAPTER SEVEN: OF MEASURELESS MAGNITUDE

1. See Walter Burkert, "Platon oder Pythagoras?: Zum Ursprung des Wortes 'Philosophie,'" *Hermes* 88.2 (1960), pp. 159–77.

2. Immanuel Kant, *Gesammelte Werke*, ed. Königlich Preußische [later Deutsche] Akademie der Wissenschaften, 23 vols. (Berlin: G. Reimer; later, De Gruyter, 1900), vol. 9, p. 28; English in *Logic*, trans. Robert S. Hartman and Wolfgang Schwarz (Indianapolis: Bobbs-Merrill, 1974), p. 31.

3. Kant, *Werke*, vol. 9, p. 28; *Logic*, p. 32.

4. Kant, *Werke*, vol. 9, p. 29; *Logic*, p. 33.

5. Kant, *Werke*, vol. 8, p. 389; English in *Raising the Tone: Late Essays by Immanuel Kant, Transformative Critique by Jacques Derrida*, ed. and trans. Peter Fenves (Baltimore: Johns Hopkins University Press, 1993), p. 51.

6. Kant, *Werke*, vol. 8, p. 392; *Raising the Tone*, p. 54.

7. Kant, *Werke*, vol. 8, pp. 392–93; *Raising the Tone*, p. 55.

8. Kant, *Werke*, vol. 8, p. 391; *Raising the Tone*, p. 53.

9. Kant, *Werke*, vol. 8, p. 393; *Raising the Tone*, p. 56. On Kant's theory of *Schwärmerei*, see Rüdiger Bubner, "Platon—der Vater aller Schwärmerei: Zu Kants Aufsatz 'Von einem neuerdings erhobenen vornehm Ton in der Philosophie,'" in *Antike Themen und ihre moderne Verwandlung* (Frankfurt am Main: Suhrkamp, 1992), pp. 80–93; Peter Fenves, "Introduction: The Topicality of Tone," in *Raising the Tone*, pp. 1–48, and, for more general considerations of the problem of "enthusiasm" and "exaltation" in critical philosophy, see Jacques Derrida, "D'un ton apocalyptique adopté naguère en philosophie," in Philippe Lacoue-Labarthe and Jean-Luc Nancy (eds.), *Les fins de l'homme: À partir du travail de Jacques Derrida* (Paris: Galilée, 1981), pp. 445–78, and Peter Fenves, "The Scale of Enthusiasm: Kant, Schelling, and Hölderlin," in *Arresting Language: From Leibniz to Benjamin* (Stanford: Stanford University Press, 2001), pp. 98–128.

10. Kant, *Werke*, vol. 8, p. 392; *Raising the Tone*, p. 55.

11. See *Kritik der Urteilskraft*, "Introduction," section 2, in *Werke*, vol. 5, pp. 174–76; English in *Critique of Judgment*, trans. Werner S. Pluhar, with a foreword by Mary J. Gregor (Indianapolis: Hackett, 1987), pp. 12–15.

12. "Introduction," section 1, in Kant, *Werke*, vol. 5, pp. 171–72; *Critique of Judgment*, pp. 9–10.

13. "Introduction," section 2, in Kant, *Werke*, vol. 5, p. 174; *Critique of Judgment*, p. 13.

14. "Introduction," section 2, in Kant, *Werke*, vol. 5, p. 175; *Critique of Judgment*, p. 14.

15. *Ibid.*

16. *Ibid.*

17. *Ibid.*

18. On the "bridge," see also "Introduction," section 10, in Kant, *Werke*, vol. 5, p. 195; *Critique of Judgment*, p. 36. On the "gulf," cf. "First Introduction," section 8, in Kant, *Werke,* vol. 5, pp. 228–29; *Critique of Judgment*, p. 418.

19. "Introduction," section 2, in Kant, *Werke*, vol. 5, pp. 175–76; *Critique of Judgment*, pp. 14–15.

20. "Preface," in Kant, *Werke*, vol. 5, p. 168; *Critique of Judgment*, p. 5.

21. *Ibid.*

22. "Introduction," section 4, in Kant, *Werke*, vol. 5, p. 179; *Critique of Judgment*, p. 19.

23. "Introduction," section 4, in Kant, *Werke*, vol. 5, pp. 179–80; *Critique of*

Judgment, p. 19.

24. "Introduction," section 4, in Kant, *Werke*, vol. 5, p. 180; *Critique of Judgment*, p. 19.

25. "Introduction," section 4, in Kant, *Werke*, vol. 5, p. 180; *Critique of Judgment*, p. 20.

26. On the terminology of *Zweckmässigkeit* in the Third *Critique*, see Giorgio Tonelli, "Von den verschiedenen Bedeutungen des Wortes Zweckmässigkeit in der Kritik der Urteilskraft," *Kant-Studien* 49 (1957–1958), pp. 154–66.

27. For a synoptic presentation, see the table in "Introduction," section 11, in Kant, *Werke*, vol. 5, p. 197; *Critique of Judgment*, p. 38.

28. "Introduction," section 6, in Kant, *Werke*, vol. 5, p. 187; *Critique of Judgment*, p. 27.

29. "Introduction," section 6, in Kant, *Werke*, vol. 5, p. 188; *Critique of Judgment*, pp. 27–28.

30. Section 15, in Kant, *Werke*, vol. 5, p. 226; *Critique of Judgment*, p. 73.

31. See section 15, in Kant, *Werke*, vol. 5, p. 226–29; *Critique of Judgment*, pp. 73–75.

32. Section 15, in Kant, *Werke*, vol. 5, p. 228; *Critique of Judgment*, p. 75.

33. "Introduction," section 7, in Kant, *Werke*, vol. 5, p. 188; *Critique of Judgment*, p. 28.

34. "Introduction," section 7, in Kant, *Werke*, vol. 5, p. 189; *Critique of Judgment*, p. 29.

35. "Introduction," section 7, in Kant, *Werke*, vol. 5, p. 190; *Critique of Judgment*, p. 30.

36. For an instance of *Harmonie*, see section 62, in Kant, *Werke*, vol. 5, p. 365; *Critique of Judgment*, p. 242.

37. Section 39, in Kant, *Werke*, vol. 5, p. 292; *Critique of Judgment*, p. 159.

38. Section 9, in Kant, *Werke*, vol. 5, p. 217–18; *Critique of Judgment*, p. 62.

39. Section 9, in Kant, *Werke*, vol. 5, p. 218; *Critique of Judgment*, p. 63.

40. See section 40, in Kant, *Werke*, vol. 5, pp. 293–96; *Critique of Judgment*, pp. 159–62.

41. On the *sensus communis*, see Jean-François Lyotard, *Leçons sur l'analytique du sublime* (Paris: Galilée, 1991), pp. 239–44.

42. See section 14, in Kant, *Werke*, vol. 5, p. 224; *Critique of Judgment*, p. 70.

43. *Ibid.*

44. See Arden Reed, "The Debt of Disinterest: Kant's Critique of Music," *MLN* 95.3 (1980), pp. 563–84, esp. 568–70; Peter Fenves, "Introduction: The Topicality of Tone," in *Raising the Tone*, esp. pp. 17–21. On Kant and music more generally see Robert E. Butts, "Kant's Theory of Musical Sound: An Early Exercise in Cognitive Science," *Dialogue* 32.1 (1993), pp. 3–24; Herbert M. Schueller, "Kant and the Aesthetics of Music," *The Journal of Aesthetics and Art Criticism* 14.2 (1955), pp. 218–47; Giselher Schubert, "Zur Musikästhetik in Kants *Kritik der Urteilskraft*," *Archiv für Musikwissenschaft* 32 (1975), pp. 12–25.

45. On the distinction between the beautiful and the agreeable, see the programmatic statement in section 3, in Kant, *Werke*, vol. 5, pp. 205–207; *Critique of Judgment*, pp. 47–48.

46. Section 53, in Kant, *Werke*, vol. 5, p. 329; *Critique of Judgment*, p. 199.

47. Section 23, in Kant, *Werke*, vol. 5, p. 247; *Critique of Judgment*, p. 100.

48. Section 23, in Kant, *Werke*, vol. 5, p. 244, *Critique of Judgment*, p. 97.

49. *Ibid.*

50. Section 23, in Kant, *Werke*, vol. 5, p. 244, *Critique of Judgment*, pp. 97–98.

51. Section 23, in Kant, *Werke*, vol. 5, p. 244; *Critique of Judgment*, p. 98.

52. On reason and the idea of totality in the sublime, see Lyotard, *Leçons sur l'Analytique du sublime*, pp. 145–53. On Kant and totality more generally, see Hans Driesch, "Kant und das Ganze," *Kant-Studien* 29 (1984), pp. 365–76.

53. Section 23, in Kant, *Werke*, vol. 5, p. 244; *Critique of Judgment*, p. 98.

54. *Ibid.*

55. Section 23, in Kant, *Werke*, vol. 5, p. 245; *Critique of Judgment*, pp. 98–99.

56. Section 23, in Kant, *Werke*, vol. 5, p. 246; *Critique of Judgment*, pp. 99–100.

57. Section 25, in Kant, *Werke*, vol. 5, p. 248; *Critique of Judgment*, p. 103.

58. *Ibid.*

59. On *magnitudo* and the terminology of quantity in Kant's works on the foundations of mathematical knowledge, see Michael Friedman, *Kant and the Exact Sciences* (Cambridge, MA: Harvard University Press, 1992), esp. "Concepts and Intuitions in the Mathematical Sciences," pp. 96–135.

60. Section 25, in Kant, *Werke*, vol. 5, p. 250; *Critique of Judgment*, p. 106.

61. *Ibid.*

62. Section 26, in Kant, *Werke*, vol. V, p. 252; *Critique of Judgment*, pp. 108–109.

63. Section 26, in Kant, *Werke*, vol. 5, p. 256; *Critique of Judgment*, p. 113.

64. Section 26, in Kant, *Werke*, vol. 5, pp. 251–52; *Critique of Judgment*,

p. 108.

65. Section 26, in Kant, *Werke*, vol. 5, p. 252; *Critique of Judgment*, p. 109.

66. Section 26, in Kant, *Werke*, vol. 5, p. 253; *Critique of Judgment*, p. 110.

67. Section 26, in Kant, *Werke*, vol. 5, p. 254; *Critique of Judgment*, p. 111.

68. Section 26, in Kant, *Werke*, vol. 5, p. 255; *Critique of Judgment*, p. 112. On the aesthetic presentation of the infinite, see Jean-Luc Nancy, "L'offrande sublime," in Jean-François Courtine et al. (eds.), *Du sublime* (Paris: Belin, 1988), pp. 37–75. On Kant and the infinite more generally, see Jonas Cohn, *Geschichte des Unendlich-keitsproblems im abendländischen Denken bis Kant* (Leipzig: W. Engelmann, 1896), pp. 231–57; Jean Seidengart, "Le traitement du problème de l'infini dans l'oeuvre de Kant," in Alain Boyer and Stéphane Chauver (eds.), *Kant analysé* (Caen: Centre de Philosophie de l'Université de Caen, 1999), pp. 115–38; A. W. Moore, "Aspects of the Infinite in Kant," *Mind* 97.386 (1988), pp. 205–23; Jacques Merleau-Ponty, *La science de l'univers à l'âge du positivisme: Étude sur les origins de la cosmologie contemporaine* (Paris: Vrin, 1983), pp. 255–73.

69. Section 26, in Kant, *Werke*, vol. 5, pp. 254–55; *Critique of Judgment*, p. 111.

70. Section 26, in Kant, *Werke*, vol. 5, pp. 255–56; *Critique of Judgment*, p. 112.

71. Section 27, in Kant, *Werke*, vol. 5, p. 257; *Critique of Judgment*, p. 114.

72. Section 27, in Kant, *Werke*, vol. 5, p. 260; *Critique of Judgment*, p. 117.

73. Section 27, in Kant, *Werke*, vol. 5, p. 258; *Critique of Judgment*, p. 115.

74. "First Introduction," section IV, in Kant, *Werke*, vol. 5, p. 209; *Critique of Judgment*, pp. 397–98.

75. *Ibid.* Cf. "Introduction," section 5, in Kant, *Werke*, vol. 5, p. 185; *Critique of Judgment*, p. 25.

CHAPTER EIGHT: OUT OF THIS WORLD

1. Letter of 19/29 August, 1599 to Michael Mästlin, in Johannes Kepler, *Gesammelte Werke*, ed. Walther von Dyck and Max Caspar (Munich: C. H. Beck, 1937–), vol. 14: *Briefe, 1599–1603*, p. 54, l.459–62. On Mästlin, see Anthony T. Grafton, "Michael Mästlin's Account of Copernican Planetary Theory," *Proceedings of the American Philosophical Society* 117.6 (1973), pp. 523–50.

2. For a summary, see Burkert, *Lore and Science in Ancient Pythagoreanism*, pp. 120–66; Kahn, *Pythagoras and the Pythagoreans*, pp. 146–53.

3. See the summaries of Pythagorean astronomy in J. L. E. Dreyer, *A History of the Planetary Systems from Thales to Kepler* (Cambridge: Cambridge University

Press, 1906), pp. 35–52; Burkert, *Lore and Science in Ancient Pythagoreanism*, pp. 299–320. In the dedicatory letter with which he opened his *De revolutionibus,* Copernicus cited Plutarch's reports of Pythagoreans according to whom the Earth revolves. See Nicholas Copernicus, *Opera omnia*, vol. 2, *De revolutionibus orbium caelestium libri sex*, ed. Ricardus Gansiniec (Warsaw: Officina Publica Libris Scientificis Edendis, 1975), pp. 4–5; English in *On the Revolutions*, ed. Jerzy Dobrzycki, trans. and commentary by Edward Rosen (Baltimore: Johns Hopkins University Press, 1978), pp. 4–5. On Copernicus and Pythagoreanism, see also Paolo Casini, "Il mito pitagorico e la rivoluzione astronomica," *Rivistia di filosofia* 85.1 (1994), pp. 7–33, esp. pp. 7–17.

4. Johannes Kepler, *Gesammelte Werke*, vol. 8, *Mysterium cosmographicum*, ed. Franz Hammer, p. 47; English in *The Secret of the Universe*, trans. A. M. Duncan, introduction and commentary by E. J. Aiton, with a preface by I. Bernard Cohen (New York: Abaris Books, 1981), p. 99.

5. Kepler, *Gesammelte Werke*, vol. 8, *Mysterium cosmographicum*, p. 23; English in *The Secret of the Universe*, p. 63.

6. See Figure 4. The illustration is in *Gesammelte Werke*, vol. 8, *Mysterium cosmographicum*, plate 1; also in *The Secret of the Universe*, p. 228.

7. Kepler, *Gesammelte Werke*, vol. 8, *Mysterium cosmographicum*, p. 45; English in *The Secret of the Universe*, pp. 93–94. In positing a basic distinction between the straight and the curved, Kepler draws, as he notes, on Nicholas Cusanus, though the mathematic theories of the two philosophers diverge on several basic points. See Dietrich Mahnke, *Unendliche Sphäre und Allmittelpunkt: Beiträge zur Genealogie der mathematischen Mystik* (Halle: M. Niemeyer, 1937), pp. 129–44, esp. 140–41.

8. Proclus in book 1 of his commentary on Euclid, I, as cited by Kepler on the title page of book 1 of his *Harmonices mundi*, in *Gesammelte Werke*, vol. 6, *Harmonices mundi*, p. 13; English in *The Harmony of the World*, trans. with introduction and notes by E. J. Aiton, A. M. Duncan, and J. V. Field (Philadelphia: American Philosophical Society, 1997), p. 7.

9. Copernicus, *Opera omnia*, vol. 2, *De revolutionibus*, p. 4; English in Copernicus, *On the Revolutions*, p. 4.

10. Copernicus, *Opera omnia*, vol. 2, *De revolutionibus*, p. 4; English in Copernicus, *On the Revolutions*, p. 4.

11. Copernicus, *Opera omnia*, vol. 2, *De Revolutionibus*, p. 4; English in Copernicus, *On the Revolutions*, p. 4. On Copernicus and symmetry, see Fernand Hallyn,

The Poetic Structure of the World: Copernicus and Kepler, trans. Donald M. Leslie (New York: Zone Books, 1993), esp. pp. 70–103.

12. Letter to Fortunio Licetti, January 1641, in *Le opere di Galileo Galilei*, ed. Antonio Favaro and Isidoro del Longo, 20 vols. (Florence: Barbera, 1890–1909), vol. 8, p. 295; cf. the better-known statement in *Il saggiatore*, section 6, in *Le opere*, vol. 6, p. 232. On Galileo and the geometrical language of nature, see Michel Blay, *Les raisons de l'infini: Du monde clos à l'univers mathémathique* (Paris: Gallimard, 1993), pp. 11–24. On Galilei and Kepler, see Massimo Bucciantini, *Galileo e Keplero: Filosofia, cosmologia, e teologia nell'età della Controriforma* (Turin: Einaudi, 2003).

13. Kepler, *Gesammelte Werke*, vol. 8, *Mysterium cosmographicum*, p. 44; English in *The Secret of the Universe*, p. 93.

14. Kepler to Michael Mästlin, October 3, 1595, in *Gesammelte Werke*, vol. 14, *Briefe, 1599–1603*, p. 35.

15. On Kepler's mathematics and the *Nova stereometria* in particular, see Dirk Jan Struik, "Kepler as a Mathematician," in *Johann Kepler, 1571–1630: A Tercentenary Commemoration of his Life and Work* (Baltimore: Williams and Wilkins, 1931), pp. 39–57, esp. pp. 44–49.

16. Kepler, *Gesammelte Werke*, vol. 6, *Harmonices mundi*, book 4, chapter 1, p. 222; English in *The Harmony of the World*, p. 302.

17. Kepler to Herwart von Hohenburg, September 14, 1599, in *Gesammelte Werke*, vol. 14, *Briefe, 1599–1603*, p. 64: "Sitque arithmetica nihil aliud…quam pars geometriae ῥητή."

18. Kepler, *Gesammelte Werke*, vol. 6, *Harmonices mundi*, book 3, preface, p. 94; English in *The Harmony of the World*, p. 131.

19. Kepler, *Gesammelte Werke*, vol. 6, *Harmonices mundi*, book 3, preface, p. 99; English in *The Harmony of the World*, p. 137.

20. *Ibid.*

21. Kepler, *Gesammelte Werke*, vol. 6, *Harmonices mundi*, book 3, preface, p. 99; English in *The Harmony of the World*, p. 138.

22. Harmonies were held to result exclusively from superparticular and multiple ratios. See Chapter 5, "Ciphers."

23. Kepler, *Gesammelte Werke*, vol. 6, *Harmonices mundi*, book 3, preface, p. 100; English in *The Harmony of the World*, p. 139.

24. See Kepler, *Gesammelte Werke*, vol. 8, *Mysterium cosmographicum*, chapter 12, pp. 66–77; English in *The Secret of the Universe*, pp. 131–37. There are reasons

to believe that Kepler himself was never fully satisfied with the chapter, as Field indicates: see J. V. Field, *Kepler's Geometrical Cosmology* (Chicago: The University of Chicago Press, 1988), pp. 112–16.

25. Kepler, *Gesammelte Werke*, vol. 6, *Harmonices mundi*, book 3, chapter 2, p. 119; English in *The Harmony of the World*, p. 164.

26. The octave is produced by the bisection of the circle.

27. See Cohen, *Quantifying Music*, pp. 20–21. His discussion owes much to D. P. Walker, "Kepler's Celestial Music," *Journal of the Warburg and Courtauld Institutes* 30 (1967), pp. 228–50, and Michael Dickreiter, *Der Musiktheoriker Johannes Kepler* (Bern: Francke Verlag, 1973), pp. 49–90.

28. It has been noted that Kepler did not foresee that in 1796, Carl Friedrich Gauss would discover that other regular polygons, such as the heptadecagon, are indeed constructible by compass and ruler. See Cohen, *Quantifying Music*, p. 21; *The Harmony of the World*, p. 145 n. 38.

29. Cohen, *Quantifying Music*, p. 21.

30. *Ibid.*

31. Kepler, *Gesammelte Werke*, vol. 6, *Harmonices mundi*, book 1, definition 7, p. 21; English in *The Harmony of the World*, p. 18.

32. Kepler, *Gesammelte Werke*, vol. 6, *Harmonices mundi*, book 3, chapter 1, axiom 7, p. 104; English in *The Harmony of the World*, p. 146. Elsewhere in the *Harmony*, Kepler again writes: "the mathematical reasons for the creation of bodies were coeternal with God." *Gesammelte Werke*, vol. 6, *Harmonices mundi*, vol. 6, book 1, chapter 1, p. 219; English in *The Harmony of the World*, p. 299. In the *Epitome*, we read: "The reasons of geometry are coeternal with God; in them one first finds the difference between the curved and the straight." *Gesammelte Werke*, vol. 7, *Epitome astronomiae Copernicae*, book 4, part 1, chapter 3, section 1, p. 267.

33. Jean-Luc Marion, *Sur la théologie blanche de Descartes: Analogie, creation des vérités éternelles et fondement* (Paris: Presses Universitaires de France, 1981), p. 179.

34. Kepler goes so far as to identify geometry with God: see *Gesammelte Werke*, vol. 6, *Harmonices mundi*, book 4, chapter 1, section 1, p. 223; English in *The Harmony of the World*, p. 304: "Why waste words? Geometry, which before the origin of things was coeternal with the divine mind and is God himself (for what could there be in God which would not be God himself?) supplied God with patterns for the creation of the world."

35. Kepler, *Gesammelte Werke*, vol. 7, *Epitome astronomiae Copernicae*, book 4,

part 2, chapter 2, section 1, p. 293.

36. Gérard Simon, *Kepler, astronome, astrologue* (Paris: Gallimard, 1979), p. 142.

37. Kepler, *Gesammelte Werke*, vol. 6, *Harmonices mundi*, book 4, chapter 1, p. 215; English in *The Harmony of the World*, p. 294.

38. *Ibid.*

39. Kepler, *Gesammelte Werke,* vol. 6, *Harmonices mundi*, book 5, chapter 3, p. 299; English in *The Harmony of the World*, p. 407. Kepler alluded to the Greek phrase in several passages. For discussions, see Silvia Tangherlini, "Temi platonici e pitagorici nell' *Harmonices mundi* di Keplero," *Rinascinmento* n.s. 14 (1974), pp. 117–78, esp. 120–24; cf. Marion, *La théologie blanche de Descartes*, pp. 191–94.

40. The formula according to which "God always geometricizes" (ὁ θεὸς ἀεὶ γεωμετρεῖ or τὸν θεὸν ἀεὶ γεωμετρεῖν) cannot be found in Plato's own works. Plutarch (*Quaestiones conviviales*, book 8, part 2, chapter 1 [718c]) attributes it to Plato while remarking upon its absence from the Platonic corpus. See Mahnke, *Unendliche Sphäre und Allmittelpunkt*, p. 140, n. 2.

41. Koyré, *From the Closed World to the Infinite Universe*, p. 2. Cf. also his *Newtonian Studies* (Cambridge: Harvard University Press, 1965), pp. 3–24.

42. Kepler, *Gesammelte Werke*, vol. 8, *Mysterium cosmographicum*, book 16, p. 92; English in *The Secret of the Universe*, p. 167. On Copernicus's views regarding the real existence of the spheres, see Nicholas Jardine, "The Significance of the Copernican Orbs," *Journal for the History of Astronomy* 13 (1982), pp. 168–94, and Peter Barker, "Copernicus, the Orbs and the Equant," *Synthese* 83 (1990), pp. 317–23.

43. Alexandre Koyré, *La révolution astronomique: Copernic, Kepler, Borelli* (Paris: Hermann, 1961), p. 194. As a caveat to his pronouncement, Koyré observes, however, that Kepler retains an exceptional realm for the fixed stars and maintains distinctions between Ptolemaic, Copernican, and Tychonic quantities. Cf. Simon, *Kepler, astronome, astrologue*, pp. 343–48.

44. Kepler, *Gesammelte Werke*, vol. 6: *Harmonices mundi*, book 5, title page, p. 287; English in *The Harmony of the World*, p. 387.

45. Kepler, *Gesammelte Werke*, vol. 6, *Harmonices mundi*, book 5, chapter 3, propositions 3 and 4, p. 300; English in *The Harmony of the World*, p. 408. See the account in Bruce Stephenson, *The Music of the Heavens: Kepler's Harmonic Astronomy* (Princeton: Princeton University Press, 1994), pp. 137–39.

46. Kepler, *Gesammelte Werke*, vol. 6, *Harmonices mundi*, book V, chapter 3,

proposition 5, p. 300; English in *The Harmony of the World*, p. 408. I have adopted Stephenson's translation in *The Music of the Heavens*, p. 136. See his helpful commentary, pp. 136–38. On this law of distance or area and the significance of the difference for Kepler, see A. J. Aiton, "Kepler's Second Law of Planetary Motion," *Isis* 60.1 (1969), pp. 75–90.

47. The principle can also be stated in the following form: "the proportion of times equals the proportion of diameters raised to the 3/2 power." Stephenson, *The Music of the Heavens*, p. 141; see also Stephenson's general discussion, *ibid.*, pp. 140–45. The Third Law is generally taught in the following form: "if T_1 and T_2 are the periods that two planets require to complete their respective orbits once, and if R_1 and R_2 are the average distances between the corresponding planets and the sun, then the ratio of the squares of the orbital periods is equal to the ratio of the cubes of the average distances from the sun, or $(T_1/T_2)^2 = (R_1/R_2)^3$." Thomas S. Kuhn, *The Copernican Revolution: Planetary Astronomy in the Development of Western Thought* (Cambridge, MA: Harvard University Press, 1957), pp. 216–17.

48. For a discussion, see Field, *Kepler's Geometrical Cosmology*, pp. 96–97.

49. Kepler, *Gesammelte Werke*, vol. 6, *Harmonices mundi*, book 5, chapter 3, p. 302; English in *The Harmony of the World*, p. 411.

50. Field, *Kepler's Geometrical Cosmology*, p. 143. Kuhn, *The Copernican Revolution*, p. 217, similarly remarks that the Third Law "did not permit astronomers to compute any quantities that were previously unknown."

51. Kepler, *Gesammelte Werke*, vol. 6, *Harmonices mundi*, book 5, preface, p. 289; English in *The Harmony of the World*, p. 389.

52. Kepler, *Gesammelte Werke*, vol. 6, *Harmonices mundi*, book 5, preface, p. 290; English in *The Harmony of the World*, p. 391.

53. Kepler, *Gesammelte Werke*, vol. 6, *Harmonices mundi*, book 5, chapter 4, p. 316; English in *The Harmony of the World*, p. 430.

54. Walker has stressed the originality of Kepler's vision of an astral polyphony. See "Kepler's Celestial Music."

55. For detailed accounts of Kepler's celestial harmonies, see Koyré, *La révolution astronomique*, pp. 328–45; Walker, "Kepler's Celestial Music"; Dickreiter, *Der Musiktheoretiker Johannes Kepler*, pp. 93–111; Owen Gingerich, "Kepler, Galilei and the Harmony of the World," in Victor Coelho (ed.), *Music and Science in the Age of Galileo* (Dordrecht: Kluwer, 1992), pp. 45–63; Stephenson, *The Music of the Heavens*, pp. 128–241.

56. Kepler, *Gesammelte Werke*, vol. 6, *Harmonices mundi*, book 5, chapter 4, p. 316; English in *The Harmony of the World*, p. 430. On Kepler's understanding of polyphony as a modern phenomenon, see Walker, "Kepler's Celestial Music," pp. 232–35.

57. See the famous terms with which the *Commentariolus* concludes in Nicolaus Copernicus, *Three Copernican Treatises: The Commentariolus of Copernicus, The Letter against Werner, The Narratio prima of Rheticus*, trans. Edward Rosen (New York: Columbia University Press, 1939), p. 90.

58. Kepler, *Gesammelte Werke*, vol. 6, *Harmonices mundi*, book 5, chapter 7, p. 328; English in *The Harmony of the World*, p. 466.

59. Hallyn, *The Poetic Structure of the World*, p. 102. Cf. the similar observation in Dickreiter, *Johannes Kepler als Musiktheoriker*, p. 98.

60. Kepler, *Gesammelte Werke*, vol. 6, *Harmonices mundi*, book 5, chapter 3, p. 303; English in *The Harmony of the World*, p. 412.

61. Koyré, *La révolution astronomique*, p. 338. Cf. Walker, "Kepler's Celestial Music," p. 249.

62. Kepler, *Gesammelte Werke*, vol. 6, *Harmonices mundi*, book 5, chapter 6, p. 322; English in *The Harmony of the World*, p. 439.

63. Walker, "Kepler's Celestial Music," pp. 247–48.

64. Dickreiter, *Johannes Kepler als Musiktheoretiker*, p. 103.

65. For Aristotle's formal definition of space, see *Physics* 4.4. 212a5–6: τὸ πέρας τοῦ περιέχοντος σώματος. Harry Austryn Wolfson, in *Crescas' Critique of Aristotle: Problems of Aristotle's Physics in Jewish and Arabic Philosophy* (Cambridge, MA: Harvard University Press, 1929), p. 116, translates: "Space is the circumambient limit of a body." Cf. Werner Hamacher, "Amphora," trans. Dana Hollander, *Assemblage* 20 (1993), pp. 40–41.

66. Aristotle, *De caelo* 1.5.271b; English in Aristotle, *On the Heavens: I and II*, ed. and trans. Stuart Leggatt (Warminster, UK: Aris and Phillips, 1995), p. 63.

67. *Ibid.*

68. *Ibid.*

69. Aristotle, *De caelo*, 1.5.273a.

70. Jean Seidengart, *Dieu, l'univers et la sphère infinie: Penser l'infinité cosmique à l'aube de la science classique* (Paris: Albin Michel, 2006), p. 104.

71. David J. Furley remarks that *mundus* and *kosmos* in Latin and Greek meant "a limited, organized system, bounded by the stars," while the universe as a whole

was called by Greek writers τὸ πᾶν, and by Latin writers (cursing their language for the absence of a definite article) "*omne quod est, omne immensum*, and so on." "The Greek Theory of the Infinite Universe," *Journal of the History of Ideas* 42.4 (1981), p. 572.

72. See Grant McColley, "The Seventeenth-Century Doctrine of the Plurality of Worlds," *Annals of Science* 1.4 (1936), p. 388.

73. Cited in McColley, "The Seventeenth-Century Doctrine," p. 389 n. 23.

74. See Nicole Oresme, *Quaestiones super Aristotelis De caelo et mundo*, in "The *Quaestiones super De celo* of Nicole Oresme*," ed. and trans. Claudia Kern, Ph.D. dissertation, University of Wisconsin, 1965 (Ann Arbor, MI: University Microfilms International, 1982), question 19, pp. 287–88.

75. *Nicole Oresme, Le Livre du ciel et du monde*, vol. 1, chapter 24, 39a–39b, p. 244.

76. Wolfson, *Crescas' Critique of Aristotle*, p. 116. For Crescas and Oresme, as well as a more recent appraisal of Crescas's physics, see Warren Zev Harvey, *Physics and Metaphysics in Hasdai Crescas* (Amsterdam: J. C. Gieben, 1998), pp. 3–45 and esp. 23–29. Cf. Warren Zev Harvey, "L'univers infini de Hasday Crescas," *Revue de métaphysique et de morale* 103.4 (1998), pp. 551–57. Cf. also Shlomo Pines, "Scholasticism after Thomas Aquinas and the Teachings of Hasdai Crescas and his Predecessors," *Proceedings of the Israel Academy of Sciences and Humanities* 1.10 (1967), rpt. in *The Collected Works of Shlomo Pines*, vol. 5: *Studies in the History of Jewish Thought*, ed. Warren Zev Harvey and Moshe Idel (Jerusalem: The Magnes Press, 1997), pp. 489–589.

77. For a helpful summary, see Seidengart, *Dieu, l'univers et la sphère infinie*, pp. 88–93.

78. Copernicus, *Opera omnia*, vol. 2, *De revolutionibus*, book 1, chapter 6, p. 14; English in *On the Revolutions*, p. 14.

79. Copernicus, *Opera omnia*, vol. 2, *De revolutionibus*, book 1, chapter 8, p. 15; English in *On the Revolutions*, p. 16. On Copernicus and the idea of an infinite universe, see Grant McColley, "Nicholas Copernicus and an Infinite Universe," *Popular Astronomy* 44.10 (1936), pp. 525–32; Seidengart, *Dieu, l'univers et la sphère infinie*, pp. 104–109.

80. See Francis R. Johnson and Sanford V. Larkey, "Thomas Digges, the Copernican System, and the Idea of the Infinity of the Universe in 1576," *The Huntington Library Bulletin* 5 (1934), pp. 69–117; Koyré, *From the Closed World to the Infinite*

Universe, pp. 28–57; Seidengart, *Dieu, l'univers et la sphère infinie*, pp. 116–41.

81. Giordano Bruno, *De immenso*, chapter 8, in *Jordani Bruni Nolani opera latine conscripta publicis sumptibus edita*, ed. Francesco Forentino and Felice Toccco, 3 vols. (Naples: D. Morano, 1879–1891), vol. 1, p. 231.

82. See *De immenso*, chapter 8, in *Jordani Bruni Nolani opera latine*, vol. 1, pp. 231–33.

83. Giordano Bruno, *De l'infinito*, Second Dialogue, in *Œuvres complètes*, ed. Yves Hersant and Nuccio Ordine, vol. 4, *De l'infini, de l'univers et des mondes*, ed. Giovanna Aquilecchia, notes by Jean Seidengart, introduction by Miguel Angel Granada, trans. Jean-Pierre Cavaillé (Paris: Belles Lettres, 1995), p. 115.

84. Letter of Edmund Bruce of November 5, 1603, in Kepler, *Gesammelte Werke*, vol. 14, *Briefe, 1599–1603*, p. 450. On the letter, its significance in representing Bruno's doctrines to Kepler, and Kepler's relation of the infinitist theses more generally, see Miguel A. Granada, "Kepler and Bruno on the Infinity of the Universe and of Solar Systems," *Journal for the History of Astronomy* 39 (2008), pp. 469–95.

85. Kepler, *Gesammelte Werke*, vol. 1, *Mysterium cosmographicum: De stella nova*, chapter 21, p. 251–52; English in Koyré, *From the Closed World to the Infinite Universe*, p. 59.

86. Kepler, *Gesammelte Werke*, vol. 1, *Mysterium cosmographicum: De stella nova*, chapter 21, p. 252–53.

87. See Aristotle, *Physics* 3.1 and 3.6, as cited by Kepler in his *De quantitatibus libelli*, chapter 6, in *Johannis Kepleri astronomi opera omnia*, ed. Christian Frisch, vol. 8, part 1 (Frankfurt am Main: Heyder and Zimmer, 1871), pp. 152–54; English in Giovanna Cifoletti, "Kepler's *De Quantitatibus*," *Annals of Science* 43 (1986), pp. 229–31.

88. Kepler, *Johannis Kepleri astronomi opera omnia*, vol. 8, part 1, p. 153; English in "Kepler's *De quantitatibus*," p. 229.

89. Kepler, *Johannis Kepleri astronomi opera omnia*, vol. 8, part 1, p. 153; English in "Kepler's *De quantitatibus*," p. 230.

90. Kepler, *Gesammelte Werke*, vol. 1, *Mysterium cosmographicum: De stella nova*, chapter 21, p. 257.

91. Kepler, *Gesammelte Werke*, vol. 1: *Mysterium cosmographicum: De stella nova*, chapter 21, p. 256.

92. For Kepler's "cosmological principle," see Field, *Kepler's Geometrical Cosmology*, p. 26. Field notes that in more modern physics, "the implied uniformity is

taken to refer not to stars, nor even, usually, to galaxies, but to clusters of galax-ies." Cf. Koyré, *From the Closed World to the Infinite Universe*, p. 78.

93. Kepler, *Gesammelte Werke*, vol. 7: *Epitome astronomiae Copernicae*, book 1, chapter 2, p. 42.

94. Kepler, *Gesammelte Werke*, vol. 8, *Mysterium cosmographicum*, p. 46; English in *The Secret of the Universe*, pp. 96–97.

95. Kepler, *Gesammelte Werke*, vol. 6, *Harmonices mundi*, preface, p. 15; English in *The Harmony of the World*, p. 9.

96. Kepler, *Gesammelte Werke*, vol. 1, *Mysterium cosmographicum: De stella nova*, chapter 21, p. 253.

97. Copernicus, *Opera omnia*, vol. 2, *De revolutionibus*, pp. 7 and 20; English in *On the Revolutions*, pp. 7 and 22.

98. Copernicus, *Opera omnia*, vol. 2, *De revolutionibus*, p. 20; English in *On the Revolutions*, p. 22.

Bibliography

A. J. Aiton, "Kepler's Second Law of Planetary Motion," *Isis* 60.1 (1969), pp. 75-90.

Alexander of Aphrodisias, *Alexandri Aphrodisiensis in Aristotelis Metaphysica commentaria*, ed. Michael Hayduck (Berlin: G. Reimer, 1891).

Apel, Willi, "Mathematics and Music in the Middle-Ages," in *Medieval Music: Collected Articles and Reviews*, forward by Thomas E. Binkley (Stuttgart: Franz Steiner Verlag, 1986), pp. 122-53.

——, *The Notation of Polyphonic Music*, 5th ed. (Cambridge: The Medieval Academy of America, 1953).

Aristotle, *On the Heavens: I and II*, ed. and trans. Stuart Leggatt (Warminster: England, 1995).

Aristoxenus, *Elementa harmonica Aristoxeni*, ed. Rosetta da Rios (Rome: Typis Publicae Officinae Polygraphicae, 1954).

Bailhache, Patrice, *Une histoire de l'acoustique musicale* (Paris: CNRS Éditions, 2001).

——, *Leibniz et la théorie de la musique* (Paris: Klincksieck, 1992).

——, "Le miroir de l'Harmonie Universelle: Musique et théorie de la musique chez Leibniz," in *L'esprit de la musique: Essais d'esthétique et de philosophie*, ed. Hugues Dufourt, Joël-Marie Fauquet, and François Hurard (Paris: Klincksieck, 1992), pp. 203-16.

——, "Le Système musical de Conrad Henfling (1706)," *Revue de musicologie* 74.1 (1981), pp. 5-25.

Barbera, C. André, "Arithmetic and Geometric Divisions of the Tetrachord, *Journal of Music Theory* 21.2 (1977), pp. 294-323.

——, *The Euclidean Division of the Canon: Greek and Latin Sources* (Lincoln:

University of Nebraska Press, 1991).

Barbour, J. Murray, "The Persistence of the Pythagorean Tuning System," *Scripta Mathematica* 1 (1933), pp. 286–304.

———, *Tuning and Temperament: A Historical Survey* (East Lansing: Michigan State University Press, 1953).

Barker, Andrew (ed. and trans.), *Greek Musical Writings*, 2 vols. (Cambridge: Cambridge University Press, 1984–1989).

———, *The Science of Harmonics in Classical Greece* (Cambridge: Cambridge University Press, 2007).

Barker, Peter, "Copernicus, the Orbs and the Equant," *Synthese* 83 (1990), pp. 317–23.

Baskevitch, François, "L'élaboration de la notion de vibration sonore: Galilée dans les *Discorsi*," *Revue d'histoire des sciences* 60.2 (2007), pp. 387–418.

Baumgarten, Alexander Gottlieb, *Aesthetica*, 3rd ed. (1750; Frankfurt an der Oder: Olms, 1986).

———, *Meditationes philosophicae de nonnullis ad poema pertinentibus*, ed. and trans. Heinz Paetzold (Hamburg: Felix Meiner, 1983).

Becker, Oskar, "Die Lehre vom Geraden und Ungeraden im Neunten Buch der Euklidischen Elemente: Versuch einer Wiederherstellung in der urpsrünglichen Gestalt," *Quellen und Studien zur Geschichte der Mathematik, Astronomie und Physik*, 3B (1936), pp. 533–53.

Belaval, Yvon, "L'idée d'harmonie chez Leibniz," in *Histoire de la philosophie: Ses problèmes, ses méthodes. Hommage à Martial Guéroult* (Paris: Fischbacher, 1964), pp. 59–78.

Benedetti, Giovanni Battista, *Diversarum speculationum mathematicarum et physicarum liber* (Taurini: apud Haeredem Bevilaquae, 1585).

Berger, Anna Maria Busse, "The Evolution of Rhythmic Notation," in *The Cambridge History of Western Music Theory*, ed. Thomas Christensen (Cambridge: Cambridge University Press, 2002), pp. 628–56.

Blay, Michel, *Les raisons de l'infini: Du monde clos à l'univers mathémathique* (Paris: Gallimard, 1993).

Boeckh, August, *Gesammelte kleine Schriften*, vol. 3, *Reden und Abhandlungen*, ed. Ferdinand Ascherson (Leipzig: B. G. Teubner, 1866).

Boethius, Anicius Manlius Severinus, *Boethian Number Theory: A Translation of the De institutione arithmetica,* ed. Michael Masi (Amsterdam: Rodopi, 1993).

——, *Fundamentals of Music*, trans. Calvin M. Bower, ed. Claude V. Palisca (New York: Yale University Press, 1989).

——, *Institution arithmétique*, ed. Jean-Yves Guillaumin (Paris: Les Belles Lettres, 1995).

——, *Traité de Musique*, ed. and trans. Christian Meyer (Turnhout: Brepols, 2004).

Boyer, Carl B., "Zero: The Symbol, the Concept, the Number," *National Mathematics Magazine* 18.8 (1944), pp. 323–30.

Bruno, Giordano, *Jordani Bruni Nolani opera latine conscripta publicis sumptibus edita*, ed. Francesco Forentino and Felice Toccco, 3 vols. (Naples: D. Morano, 1879–1891).

Bubner, Rüdiger, "Platon—der Vater aller Schwärmerei: Zu Kants Aufsatz 'Von einem neuerdings erhobenen vornehm Ton in der Philosophie,'" in *Antike Themen und ihre moderne Verwandlung* (Frankfurt am Main: Suhrkamp, 1992), pp. 80–93.

Bucciantini, Massimo, *Galileo e Keplero: Filosofia, cosmologia, e teologia nell'età della Controriforma* (Turin: Einaudi, 2003).

Burkert, Walter, *Lore and Science in Ancient Pythagoreanism*, trans. Edwin L. Minar (Cambridge, MA: Harvard University Press 1972).

——, "Platon oder Pythagoras?: Zum Ursprung des Wortes 'Philosophie,'" *Hermes* 88.2 (1960), pp. 159–77.

Butts, Robert E., "Kant's Theory of Musical Sound: An Early Exercise in Cognitive Science," *Dialogue* 32.1 (1993), pp. 3–24.

Casini, Paolo, "Il mito pitagorico e la rivoluzione astronomica," *Rivistia di filosofia* 85.1 (1994), pp. 7–33.

Clagett, Marshall, *Giovanni Marliani and Late Medieval Physics* (New York: Columbia University Press, 1941).

——, *The Science of Mechanics in the Middle Ages* (Madison: The University of Wisconsin Press, 1959).

Cohen, David E., "Notes, Scales, and Modes in the Earlier Middle Ages," in *The Cambridge History of Western Music Theory*, ed. Thomas Christensen (Cambridge: Cambridge University Press, 2002), pp. 307–63.

Cohen, H. F., *Quantifying Music: The Science of Music at the First Stage of the Scientific Revolution, 1580–1650* (Dordrecht: D. Reidel, 1984).

Cohn, Jonas, *Geschichte des Unendlichkeitsproblems im abendländischen Denken bis Kant* (Leipzig: W. Engelmann, 1896).

Copernicus, Nicholas, *On the Revolutions*, ed. Jerzy Dobrzycki, trans. and commentary by Edward Rosen (Baltimore: Johns Hopkins University Press, 1978).

——, *Opera omnia*, 2 vols. (Warsaw: Officina Publica Libris Scientificis Edendis, 1975).

Curtius, Ernst Robert, *European Literature and the Latin Middle Ages*, trans. Willard R. Trask (Princeton: Princeton University Press, 1953).

Dalla Chiara, Maria Luisa, Roberto Giuntini, and Federico Laudisa (eds.), *Language, Quantum, Music: Selected Contributed Papers of the Tenth International Congress of Logic, Methodology and Philosophy of Science, Florence, August 1995* (Dordrecht: Kluwer, 1990).

Della Seta, Fabrizio, "Idee musicali nel *Tractatus de configurationibus qualitatum et motuum* di Nicola Oresme," in *La musica nel tempo di Dante: Atti del convegno internazionale, Ravenna, 12–14 settembre 1986* (Milan: Unicopli, 1988), pp. 222–56.

Delatte, Armand, *Études sur la littérature pythagoricienne* (Paris: Champion, 1915).

Derrida, Jacques, "D'un ton apocalyptique adopté naguère en philosophie," in Philippe Lacoue-Labarthe and Jean-Luc Nancy (eds.), *Les fins de l'homme: À partir du travail de Jacques Derrida* (Paris: Galilée, 1981), pp. 445–78.

Dickreiter, Michael, *Der Musiktheoriker Johannes Kepler* (Bern: Francke Verlag, 1973).

Diels, Hermann, and Walter Kranz, *Die Fragmente der Vorsokratiker*, 7th ed. (Berlin: Weidmann, 1954).

Drake, Stillman, "Bradwardine's Function, Mediate Denomination and Multiple Continua," *Physis* 12 (1970), pp. 51–68.

——, "Renaissance Music and Experimental Science," *Journal of the History of Ideas* 31.4 (1970), pp. 483–500.

Driesch, Hans, "Kant und das Ganze," *Kant-Studien* 29 (1984), pp. 365–76.

Duchez, Marie-Elisabeth, "Des neumes à la portée: Élaboration et organization rationnelles de la discontinuité musicale et sa representation graphique, de la formule mélodique à l'échelle monocordale," *Revue de musique des universités canadiennes* 4 (1983), pp. 22–65.

Dufourt, Hugues, *Essai sur les principes de la musique,* vol. I, *Mathêsis et subjectivité: Des conditions historiques de possibilité de la musique occidentale* (Paris: Musica Ficta, 2007).

Euclid, *The Thirteen Books of Euclid's Elements*, trans. Sir Thomas Heath, 2nd ed., 3 vols. (Cambridge: Cambridge University Press, 1956).

Fenves, Peter, *Arresting Language: From Leibniz to Benjamin* (Stanford: Stanford University Press, 2001).

Field, J. V., *Kepler's Geometrical Cosmology* (Chicago: University of Chicago Press, 1988).

Friedman, Michael, *Kant and the Exact Sciences* (Cambridge, MA: Harvard University Press, 1992).

Furley, David J., "The Greek Theory of the Infinite Universe," *Journal of the History of Ideas* 42.4 (1981), pp. 571–85.

Gaffurius, Franchinus, *Theorica musicae*, ed. Ilde Illuminati and Cesarino Ruini, trans. Ilde Illuminati, with an essay by Fabio Bellissima (Florence: Edizioni del Galluzzo, 2005).

Galilei, Galileo, *Le opere di Galileo Galilei*, ed. Antonio Favaro and Isidoro del Longo, 20 vols. (Florence: Barbera, 1890–1909).

——, *Opere*, ed. Franz Brunetti, 2 vols. (Turin: UTET, 1964).

Galilei, Vincenzo, *Discorso intorno alle opere di Gioseffo Zarlino et altri importanti particolari attenenti alla musica* (Venice: n.p., 1589).

Gardiès, Jean-Louis, *L'héritage épistémologique d'Eudoxe de Cnide* (Paris: Vrin, 1968).

Gingerich, Owen, "Kepler, Galilei and the Harmony of the World," in Victor Coelho (ed.), *Music and Science in the Age of Galileo* (Dordrecht: Kluwer, 1992), pp. 45–63.

Grafton, Anthony T., "Michael Mästlin's Account of Copernican Planetary Theory," *Proceedings of the American Philosophical Society* 117.6 (1973), pp. 523–50.

Granada, Miguel A., "Kepler and Bruno on the Infinity of the Universe and of Solar Systems," *Journal for the History of Astronomy* 39 (2008), pp. 469–95.

Guillaumin, Jean-Yves, "Le terme quadrivium de Boèce et ses aspects moraux," *L'Antiquité Classique* 59 (1990), pp. 139–48.

Hallyn, Fernand, *The Poetic Structure of the World: Copernicus and Kepler*, trans. Donald M. Leslie (New York: Zone Books, 1993).

Hamacher, Werner, "Amphora," trans. Dana Hollander, *Assemblage* 20 (1993), pp. 40–41.

Harvey, Warren Zev, *Physics and Metaphysics in Hasdai Crescas* (Amsterdam: J. C. Gieben, 1998).

——, "L'univers infini de Hasday Crescas," *Revue de métaphysique et de morale* 103.4

(1998), pp. 551–57.

Heller, Siegfried, "Die Entdeckung der stetige Teilung durch die Pythagoreer," *Abhandlungen der Deutschen Akademie der Wissenschaften zu Berlin, Klasse für Mathematik, Physik und Technik* 6 (1958), pp. 5–28.

Heath, Sir Thomas, *A History of Greek Mathematics*, 3 vols. (Oxford: Clarendon Press, 1921).

Huffman, Carl, *Archytas of Tarentum: Pythagorean, Philosopher, Mathematician King* (Cambridge: Cambridge University Press, 2005).

——, "The Role of Number in Philolaus' Philosophy," *Phronesis* 33.1 (1988), pp. 1–30.

——, *Philolaus of Croton: Pythagorean and Presocratic: A Commentary on the Fragments and Testimonia with Interpretative Essays* (Cambridge: Cambridge University Press, 1993).

Iamblichus, *De vita Pythagorica Liber* ed. Ludwig Deubner (Lepzig: G. B. Teuber, 1937).

Jan, Karl von (ed.), *Musici scriptores graeci: Aristoteles, Euclides, Nicomachus, Bacchus, Gaudentius, Alypius et melodiarum veterum quidquid extat*, 2 vols. (Leipzig: G. B. Teubner, 1895).

Jardine, Nicholas, "The Significance of the Copernican Orbs," *Journal for the History of Astronomy* 13 (1982), pp. 168-194.

Johnson, Francis R., and Sanford V. Larkey, "Thomas Digges, the Copernican System, and the Idea of the Infinity o fthe Universe in 1576," *The Huntington Library Bulletin* 5 (1934), pp. 69–117.

Junge, Gustav, "Wann haben die Griechen das Irrationale entdeckt?" in *Novae Symbolae Joachimicae: Festschrift des königlichen Joachimsthalschen Gymnasium* (Halle: Verlag der Buchhandlung des Waisenhauses, 1907), pp. 221–66.

Kahn, Charles H., *The Art and Thought of Heraclitus: An Edition of the Fragments with Translation and Commentary* (Cambridge: Cambridge University Press, 1979).

——, *Pythagoras and the Pythagoreans: A Brief History* (Indianapolis: Hackett, 2001).

Kant, Immanuel, *Critique of Judgment*, trans. Werner S. Pluhar, with a foreward by Mary J. Gregor (Indianapolis: Hackett, 1987).

——, *Gesammelte Werke*, ed. Königlich Preußische [later Deutsche] Akademie der Wissenschaften (Berlin: G. Reimer; later, De Gruyter, 1900–).

——, *Logic*, trans. Robert S. Hartman and Wolfgang Schwarz (Indianapolis: Bobbs-Merill, 1974).

——, *Raising the Tone: Late Essays by Immanuel Kant, Transformative Critique by Jacques Derrida*, ed. and trans. Peter Fenves (Baltimore: Johns Hopkins University Press, 1993).

Kepler, Johannes, *Gesammelte Werke*, ed. Walther von Dyck and Max Caspar (Munich: C. H. Beck, 1937–).

——, *The Harmony of the World*, trans. with introduction and notes by E. J. Aiton, A. M. Duncan, and J. V. Field (Philadelphia: American Philosophical Society, 1997).

——, *Johannis Kepleri astronomi opera omnia*, ed. Christian Frisch, 8 vols. (Frankfurt am Main: Heyder and Zimmer, 1871).

——, "Kepler's *De Quantitatibus*," ed. and trans. Giovanna Cifoletti, *Annals of Science* 43 (1986), pp. 213–38.

——, *The Secret of the Universe*, trans. A. M. Duncan, introduction and commentary by E. J. Aiton, with a preface by I. Bernard Cohen (New York: Abaris Books, 1981).

Kittler, Friedrich, *Musik und Mathematik*, part 1, *Hellas*, 2 vols. (Munich: Wilhelm Fink, 2006).

Klein, Jacob, *Greek Mathematical Thought and the Origins of Algebra*, trans. Eva Brann (Cambridge, MA: The MIT Press, 1968).

Knorr, Wilbur Richard, *The Evolution of the Euclidean Elements: A Study of the Theory of Incommensurable Magnitudes and Its Significance for Early Greek Geometry* (Dordrecht: Kluwer, 1975).

Koyré, Alexandre, *Études d'histoire de la pensée philosophique* (Paris: Armand Colin, 1961).

——, *From the Closed World to the Infinite Universe* (Baltimore: Johns Hopkins University Press, 1957.

——, *Newtonian Studies* (Cambridge, MA: Harvard University Press, 1965).

——, Review of Anneliese Maier, *Die Vorläufer Galileis im 14. Jahrhundert: Studien zur Naturphilosophie der Spätscholastik* (Rome: Edizioni Storia e letteratura, 1949), *Archives internationales d'histoire des sciences*, n. s. d'Archeion 4:14, 30 (1951), pp. 769–83.

——, *La révolution astronomique: Copernic, Kepler, Borelli* (Paris: Hermann, 1961).

Kucharski, Paul, *Étude sur la doctrine pythagoricienne de la tétrade* (Paris: Les Belles Lettres, 1952).

Kuhn, Thomas S. *The Copernican Revolution: Planetary Astronomy in the Development *of Western Thought* (Cambridge, MA: Harvard University Press, 1957).

Lasserre, François, *La naissance des mathématiques à l'époque de Platon* (Paris: Éditions du Cerf, 1990).

Leibniz, Gottfried Wilhelm, *Der Briefwechsel zwischen Leibniz und Conrad Henfling*, ed. Rudolf Haase (Frankfurt am Main: Vittorio Klostermann, 1982).

———, *Epistolae ad diversos theologici, iuridici, medici, philosophici, mathematici, historici et philologici argumenti e msc. auctoris cum annotationibus suis primum divulgavit*, ed. Christian Kortholt, 2 vols. (Leipzig: 1734).

———, *Philosophical Essays*, trans. Roger Ariew and Daniel Garber (Indianapolis: Hackett, 1989).

———, *Die philosophische Schriften von Gottfried Wilhelm Leibniz*, ed. C. J. Gerhardt, 7 vols. (1875–1890; Hildesheim: Olms, 1978).

Levy, Kenneth, "On the Origin of Neumes," *Early Music History* 7 (1987), pp. 59–90.

Lindley, Mark, s.v. "Temperament," in Stanley Sadie (ed.), *The New Grove Dictionary of Music and Musicians*, 20 vols. (London: Macmillan, 1980).

Lohmann, Johannes, *Mousiké und Logos: Aufsätze zur griechischen Philosophie und Musiktheorie, zum 75. Geburtstag des Verfassers am 9 Juli 1970*, ed. Anastasios Giannarás (Stuttgart: Musikwissenschaftliche Verlagsgesellschaft, 1970).

Luppi, Andrea, *Lo specchio dell'armonia universale: Estetica e musica in Leibniz* (Milan: Franco Angeli, 1989).

Lyotard, Jean-François, *Leçons sur l'analytique du sublime* (Paris: Galilée, 1991).

Macrobius, Ambrosius Theodosius, *Ambrosii Theodosii Macrobii Commentarii in somnium Scipionis*, ed. Jacob Willis (Leipzig: G. B. Teubner, 1963).

———, *Commentary on the Dream of Scipio*, ed. William Harris Stahl (New York: Columbia University Press, 1952).

Mahnke, Dietrich, *Unendliche Sphäre und Allmittelpunkt: Beiträge zur Genealogie der mathematischen Mystik* (Halle: M. Niemeyer, 1937).

Maier, Anneliese, *Die Vorläufer Galileis im 14. Jahrhundert*, 2nd ed. (Rome: Storia e letteratura, 1966).

———, *Zwei Grundprobeme der scholastischen Naturphilosophie: Das Problem der intensiven Große; Die Impetustheorie*, 3rd ed. (Rome: Edizioni di Storia e letteratura, 1968).

Marion, Jean-Luc, *Sur la théologie blanche de Descartes: Analogie, création des vérités éternelles et fondement* (Paris: Presses Universitaires de France, 1981).

Mathiesen, Thomas J., "An Annotated Translation of Euclid's 'Division of the Monochord,'" *Journal of Music Theory* 19.2 (1975), pp. 236–58.

McColley, Grant, "The Seventeenth-Century Doctrine of the Plurality of Worlds," *Annals of Science* 1.4 (1936), pp. 385–430.

McVaugh, Michael, "Arnald of Villanova and Bradwardine's Law," *Isis* 58.1 (1967), pp. 56–64.

Merleau-Ponty, Jacques, *La science de l'univers à l'âge du positivisme: Étude sur les origins de la cosmologie contemporaine* (Paris: Vrin, 1983).

Meyer, P. Bonaventura, APMONIA: *Bedeutungsgeschichte des Worte von Homer bis Aristoteles* (Zurich: A.-G. Gebr. Leemann & Co., 1925).

Milner, Jean-Claude, *L'œuvre claire: Lacan, la science, la philosophie* (Paris: Seuil, 1995).

Mondadori, Fabrizio, "A Harmony of One's Own and Universal Harmony in Leibniz's Paris Writings," in *Leibniz à Paris (1672–1676), Symposium de la Leibniz-Gesellschaft (Hannover) et du Centre National de la Recherche Scientifique (Paris) à Chantilly (France) du 14 au 18 novembre, 1976*, vol. 2, *La Philosophie de Leibniz* (Wiesbaden: F. Steiner, 1978), pp. 151–68.

Moore, A. W., "Aspects of the Infinite in Kant," *Mind* 97.386 (1988), pp. 205–23.

Moutsopoulos, Evanghélos, *La musique dans l'œuvre de Platon* (Paris: Presses Universitaires de France, 1959).

Murdoch, John E., "The Medieval Language of Proportions: Elements of the Interaction with Greek Foundations and the Development of New Mathematical Techniques," in A. C. Crombie (ed.), *Scientific Change: Historical Studies in the Intellectual, Social and Technical Conditions for Scientific Discovery and Technical Invention, from Antiquity to the Present* (New York: Basic Books, 1963), pp. 237–71.

Nancy, Jean-Luc, "L'offrande sublime," in Jean-François Courtine et al. (eds.), *Du sublime* (Paris: Belin, 1988), pp. 37–75.

Nicomachus of Gerasa, *In Nicomachi Arithmeticam introductionem liber*, ed. Hermenegildus Pistelli (Leipzig: B. G. Teubner, 1894).

———, *The Manual of Harmonics of Nicomachus the Pythagorean*, ed. and trans. Flora R. Levin (Grand Rapids, FL: Phanes Press, 1994).

Olimpiodorus, "Olympiodori Philosophi Scholia in Platonis Gorgiam," ed. Albert Jahn, *Neue Jahrbücher für Philologie und Pädagogik, oder Kritische Bibliothek für das Schul- und Unterrichtswesen*, 14.1 (1848), pp. 104–49.

Oresme, Nicole, *Nicole Oresme and the Kinematics of Circular Motion: Tractatus de commensurabilitate vel incommensurabilitate motuum celi*, ed. and trans. Edward Grant (Madison: University of Wisconsin Press, 1971).

——,*Nicole Oresme, Le livre du ciel et du monde*, ed. Albert D. Menut and Alexander J. Denomy, trans. Albert D. Menut (Madison: University of Wisconsin Press, 1968).

——, *Nicole Oresme and the Medieval Geometry of Qualities and Motions: A Treatise on the Uniformity and Difformity of Intensities Known as Tractatus de configurationibus qualitatum et motuum*, ed. and trans. Marshall Clagett (Madison: University of Wisconsin Press, 1968).

——, *Nicole Oresme, De proportionibus proportionum and Ad pauca respicientes*, ed. Edward Grant (Madison: University of Wisconsin Press, 1966).

——, *Quaestiones super Aristotelis De caelo et mundo*, in "The Quaestiones super De celo of Nicole Oresme," ed. and trans. Claudia Kern, Ph.D. dissertation, University of Wisconsin, 1965 (Ann Arbor, MI: University Microfilms International, 1982).

Palisca, Claude V., "Scientific Empiricism in Musical Thought," in Hedley Howell Rhys (ed.), *Seventeenth Century Science and the Arts* (Princeton: Princeton University Press, 1961), pp. 91–137.

Pappus, *The Commentary of Pappus on Book X of Euclid's Elements: Arabic Text and Translation*, ed. William Thomson and Gustav Junge (Cambridge, MA: Harvard University Press, 1930).

Pines, Shlomo, *The Collected Works of Shlomo Pines*, vol. 5, *Studies in the History of Jewish Thought*, ed. Warren Zev Harvey and Moshe Idel (Jerusalem: Magnes Press, 1997).

Plato, *Platonis Dialogi secundum Thrasylli tetralogias dispositi*, ed. Karl Friedrich Hermann (Leipzig: B. G. Teubner, 1927).

Proclus, *Procli Diadochi in primum Euclidis Elementorum librum commentarii*, ed. Gottfried Friedlein (Leipzig: B. G. Teubner, 1873).

Reed, Arden, "The Debt of Disinterest: Kant's Critique of Music," *MLN* 95.3 (1980), pp. 563–84.

Roger Bacon, *Opera hactenus inedita*, ed. Robert Steele, 16 vols. (Oxford: Clarendon Press, 1905–).

Rosen, Edward (ed. and trans.), *Three Copernican Treatises: The Commentariolus of Copernicus, The Letter against Werner, The Narratio prima of Rheticus* (New York:

Columbia Universiy Press, 1939).

Scholz, Heinrich, "Warum haben die Griechen die Irrationalzahlen nicht aufgebaut?" *Kant-Studien* 33 (1928), pp. 35–72.

Schubert, Giselher, "Zur Musikästhetik in Kants *Kritik der Urteilskraft*," *Archiv für Musikwissenschaft* 32 (1975), pp. 12–25.

Schueller, Herbert M., "Kant and the Aesthetics of Music," *The Journal of Aesthetics and Art Criticism* 14.2 (1955), pp. 218–47.

Seidengart, Jean, *Dieu, l'univers et la sphère infinie: Penser l'infinité cosmique à l'aube de la science classique* (Paris: Albin Michel, 2006).

———, "Le traitement du problème de l'infini dans l'œuvre de Kant," in *Kant analysé*, ed. Alain Boyer and Stéphane Chauver (Caen: Centre de Philosophie de l'Université de Caen, 1999), pp. 115–38.

Simon, Gérard, *Kepler, astronome, astrologue* (Paris: Gallimard, 1979).

Spitzer, Leo, *Classical and Christian Ideas of World Harmony: Prolegomenon to an Interpretation of the Word 'Stimmung,'* ed. Anna Granville Hatcher, preface by René Wellek (Baltimore: Johns Hopkins University Press, 1963).

Stephenson, Bruce, *The Music of the Heavens: Kepler's Harmonic Astronomy* (Princeton: Princeton University Press, 1994).

Stevin, Simon, *Les œuvres mathématiques de Simon Stevin de Bruges* (Leiden: n. p. 1634).

———, *The Principal Works of Simon Stevin*, ed. Ernst Crone, E. J. Diksterhuis, R. J. Forbes, M. G. Minnaert, and A. Pannekoek, 5 vols. In 6 (Amsterdam: Swets and Zeitlinger, 1955–).

Struik, Dirk Jan, "Kepler as a Mathematician," in *Johann Kepler, 1571–1630: A Tercentenary Commemoration of his Life and Work* (Baltimore: Williams and Wilkins, 1931), pp. 39–57.

Sylla, Edith D., "Medieval Concepts of the Latitude of Forms: The Oxford Calculators," *Archives d'histoire doctrinale et littéraire du Moyen Âge* 40 (1973), pp. 223–83.

Szabó, Árpád, *Anfänge der griechischen Mathematik* (Budapest: Akadémiai Kiádo, 1969).

———, *Die Entfaltung der griechischen Mathematik* (Leipzig: Bibliographisches Institut, 1994).

Tangherlini, Silvia, "Temi platonici e pitagorici nell'*Harmonice mundi* di Keplero," *Rinascinmento* n.s. 14 (1974), pp. 117–78.

Tannery, Paul, *Mémoires scientifiques*, ed. J. L. Heiberg and H. G. Zeuthen, 7 vols. (Paris: E. Privat, 1912–25).

Taschow, Ulrich, "Die Bedeutung der Musik als Modell für Nicole Oresmes Theorie: *De configurationibus qualitatum et motuum*," *Early Science and Medecine* 4.1 (1999), pp. 37–90.

Theodosius of Bithynia, *Theodosii De habitationibus; De diebus et noctibus, Abhandlungen der Gesellschaft der Wissenschaften zu Göttingen*, ed. and trans. Rudolf Fecht (Berlin: 1927).

Theon of Smyrna, *Theonis Smyrnaei, Philosophi platonici: Expositio rerum mathematicarum ad legendum Platonem utilium*, ed. Eduard Hiller (Leipzig: B. G. Teubner, 1878).

Theophrastus of Eresus, *Theophrastus of Eresus: Sources for His Life, Writings, Thought and Influence*, ed. and trans. William W. Fortenbaugh, Pamela Huby, Robert W. Sharples, and Dimitri Gutas, 8 vols. (Leiden: E. J. Brill, 1992–2007).

Thomas of Bradwardine, *Thomas of Bradwardine: His Tractatus de Proportionibus*, ed. Lamar Crosby, Jr. (Madison: The University of Wisconsin Press, 1955).

Tonelli, Giorgio, "Von den verschiedenen Bedeutungen des Wortes Zweckmässigkeit in der Kritik der Urteilskraft," *Kant-Studien* 49 (1957–1958), pp. 154–66.

Van der Waerden, B. L., "Arithmetik der Pythagoreer," *Mathematische Annalen* 120 (1947–1949), pp. 127–53.

Vogt, Heinrich, "Die Entdeckungsgeschichte des Irrationalen nach Plato und anderen Quellen des 4. Jahrhunderts," *Bibliotheca mathematica* 10 (1909–1910), pp. 97–15.

Von Fritz, Kurt, "The Discovery of Incommensurability by Hippasus of Metapontum," in David J. Furley and R. E. Allen (eds.), *Studies in Presocratic Philosophy*, 2 vols. (New York: Humanities Press, 1970–1975), 2 vols., vol. 1, pp. 409–12.

Walker, D. P., "Kepler's Celestial Music," *Journal of the Warburg and Courtauld Institutes* 30 (1967), pp. 228–50.

——, "Musical Humanism in the 16th and Early 17th Centuries," *The Music Review* 2 (1941), pp. 1–13; *The Music Review* 2.2, pp. 111–21; *The Music Review* 2.3 (1942), pp. 220–27; *The Music Review* 2.4 (1942), pp. 288–308; *The Music Review* 3 (1942), pp. 55–71.

——, "Some Aspects of the Musical Theory of Vincenzo Galilei and Galileo Galilei," *Proceedings of the Royal Musical Association* 100 (1973–1974), pp. 33–47.

Wallis, John, *Opera mathematica*, 3 vols. (Oxoniae, e theatro Sheldoniano,

1693–1699).

Weil, Simone, *Sur la science* (Paris: Gallimard, 1966).

Werckmeister, Andreas, *Musicalische Temperatur*, ed. Rudolf Rasch (Utrecht: Diapason Press, 1983).

Wersinger, Anne Gabrièle, *La sphère et l'intervalle: Le schème de l'Harmonie dans la pensée des anciens Grecs d'Homère à Platon* (Paris: Jerôme Millon, 2008).

West, M. L., *Ancient Greek Music* (Oxford: Oxford University Press, 1992).

Wienpahl, Robert W., "Zarlino, the Senario, and Tonality," *Journal of the American Musicological Society*, 12.1 (1959), pp. 27–41.

Winnington-Ingram, R. P., "Aristoxenus and the Intervals of Greek Music," *The Classical Quarterly* 26:3–4 (1932), pp. 195–208.

Wolfson, Harry Austryn, *Crescas' Critique of Aristotle: Problems of Aristotle's Physics in Jewish and Arabic Philosophy* (Cambridge, MA: Harvard University Press, 1929).

Zarlino, Gioseffe, *Instituzioni armoniche* (Venice: n. p. 1558).

Zedda, Sergio, "How to Build a World Soul: A Practical Guide," in M. R. Wright (ed.), *Reason and Necessity: Essays on Plato's Timaeus* (London: Duckworth, 2000), pp. 23–41.

Zeuthen, Hieronymus Georg, "Sur l'origine historique de la connaissance des quantités irrationelles," *Oversigt over det Koneglige Danske videnskabernes Selskabs Ferhandlinger* (1915), pp. 333–52.

Zhmud', Leonid Ja., "'All is Number'?: 'Basic Doctrine' of Pythagoreanism Reconsidered," *Phronesis* 34.3 (1989), pp. 270–92.

Zoubov, V., "Nicole Oresme et la musique," *Medieval and Renaissance Studies* 5 (1961), pp. 96–10.

Index

Zone Books series design by Bruce Mau
Typesetting by Meighan Gale
Image placement and production by Julie Fry
Printed and bound by Thomson-Shore